Robert Bentley

A Text-Book of Organic Materia Medica

Robert Bentley

A Text-Book of Organic Materia Medica

ISBN/EAN: 9783337371098

Printed in Europe, USA, Canada, Australia, Japan

Cover: Foto ©berggeist007 / pixelio.de

More available books at **www.hansebooks.com**

A TEXT-BOOK

OF

ORGANIC MATERIA MEDICA

COMPRISING A DESCRIPTION OF THE

VEGETABLE AND ANIMAL DRUGS

OF THE

British Pharmacopœia

WITH OTHER NON-OFFICIAL MEDICINES

ARRANGED SYSTEMATICALLY AND ESPECIALLY DESIGNED FOR THE USE OF STUDENTS

BY

ROBERT BENTLEY, M.R.C.S.Eng., F.L.S.

FELLOW OF KING'S COLLEGE, LONDON ; HONORARY MEMBER OF THE PHARMACEUTICAL
SOCIETY OF GREAT BRITAIN ; HONORARY MEMBER OF THE AMERICAN PHARMACEUTICAL
ASSOCIATION ; HONORARY MEMBER OF THE PHILADELPHIA COLLEGE OF PHARMACY ;
MEDICAL ASSOCIATE OF KING'S COLLEGE, LONDON ; PROFESSOR OF BOTANY
AND MATERIA MEDICA TO THE PHARMACEUTICAL SOCIETY OF GREAT
BRITAIN ; PROFESSOR OF BOTANY IN KING'S COLLEGE, LONDON ; ONE
OF THE THREE EDITORS OF THE 'BRITISH PHARMACOPŒIA' 1885

WITH MANY ILLUSTRATIONS ON WOOD

LONDON

LONGMANS, GREEN, AND CO.

1887

TO

MICHAEL CARTEIGHE, F.I.C., F.C.S., &c.

PRESIDENT OF THE PHARMACEUTICAL SOCIETY OF GREAT BRITAIN

THIS WORK IS DEDICATED

AS A SMALL TRIBUTE OF ACKNOWLEDGMENT OF HIS UNTIRING ENERGY

ZEAL AND ABILITY IN THE PROMOTION AND DEVELOPMENT OF

PHARMACEUTICAL EDUCATION AND EXAMINATION

AND

WITH EVERY SENTIMENT OF PERSONAL REGARD AND ESTEEM

BY HIS SINCERE FRIEND

THE AUTHOR

PREFACE.

THE Author has, for many years past, contemplated the
compilation of such a work as the present, for the use of
students during their apprenticeship, and as a text-book for
them while attending courses of lectures, and as a preparation for their examinations With these special objects in
view, the Author has endeavoured to make it as elementary
as possible, but, at the same time, thoroughly practical and in
accordance with the present state of the science on which it
treats. The general characters of the various drugs derived
from the Animal and Vegetable Kingdoms have been given
very fully, so as to enable the student to recognise them with
facility and certainty, and thus at the same time readily to
detect any adulteration of the genuine drug, or the substitution of the false for the true.   In this respect the Author
believes that the present work will not only be a trustworthy
guide to the student, but also especially valuable to the
pharmacist generally, and to all engaged in the prescribing
and dispensing of medicines.

It will be noticed that the arrangement of the plants
from which the Vegetable Drugs are derived is different
in many respects from that ordinarily adopted in works on
Materia Medica. This arrangement is founded, so far as
the Phanerogamia are concerned, upon that adopted by
Bentham and Hooker in their 'Genera Plantarum,' which
great work cannot fail to be the standard authority on

the subject for many years to come, and the arrangement there adopted must consequently, in this country, at least, come into general use. Indeed, the Author hoped to have brought out the present work immediately after the publication of the British Pharmacopœia in 1885, but was induced to defer it until he had fully explained the arrangement in the fifth edition recently issued of his ' Manual of Botany.'

The Table of Contents in the present work has been fully and systematically arranged, with the view of bringing before the student at a glance the sources of the various drugs, as well as the orders and higher divisions of the Vegetable and Animal Kingdoms to which the plants and animals yielding them respectively belong, more especially those of vegetable origin, which constitute by far the larger proportion of the articles of the Materia Medica. The Author would therefore recommend that, before commencing the study of the drugs of any particular order, reference should be first made to the Table of Contents, when the position of the order will be seen, the names and number of medicinal plants treated of which it contains, and the parts and products of each which are used in medicine. By studying in this way the student will acquire not only a special knowledge of the several drugs, but also a general acquaintance with the several groups in which they are arranged and classified.

LONDON : *April*, 1887.

CONTENTS.

—◦✦◦—

PART I.

VEGETABLE DRUGS.

SUB-KINGDOM I.

PHANEROGAMIA OR FLOWERING PLANTS.

DIVISION I. *ANGIOSPERMIA.*

CLASS I. DICOTYLEDONES.

SUB-CLASS I. POLYPETALÆ.

SERIES I.—*THALAMIFLORÆ.*

Contents.   xi

xii   *Contents.*

Contents. xiii

Contents.

xvi *Contents.*

a

a 2

Contents.   xxi

SUB-KINGDOM II.

CRYPTOGAMIA OR FLOWERLESS PLANTS.

DIVISION I. *CORMOPHYTA.*

DIVISION II. *THALLOPHYTA.*

PART II.

ANIMAL DRUGS.

SUB-KINGDOM I.

VERTEBRATA.

CLASS I. MAMMALIA.

Contents.   xxvii

CLASS II. ANNELIDA OR ANNULOSA.

CLASS III. PORIFERA OR SPONGIDA.

THE

STUDENT'S TEXT-BOOK

OF

ORGANIC MATERIA MEDICA.

———◦◦———

GENERAL INTRODUCTION.

MATERIA MEDICA, in the limited sense indicated by the term, is that branch of medical science which comprises the consideration of all material substances employed as medicines or remedies, or, in other words, as curative agents in the treatment of disease. But by many writers it has a far wider range, and includes a notice of all kinds of remedies, from whatever source derived—that is, whether psychical or mental, imponderable, hygienic, mechanical, surgical, or pharmacological.

In this volume we limit its application to pharmacological remedies, or those material substances which are capable of modifying or altering the vital actions of the body, and thus to act as curative agents in the treatment of disease. Materia Medica may, therefore, be also termed Pharmacology (from φάρμακον, *a medicine*, and λόγος, *a discourse*).

Materia Medica or Pharmacology has been subdivided into the following three departments : Pharmacognosy, Pharmacy, and Therapeutics.

Pharmacognosy (from φάρμακον, *a medicine*, and γι-γνώσκω, *I know*) has reference to the source, characters,

B

varieties, chemical composition, and purity of *raw* or *un-prepared medicines*, or *simples* as they are also frequently termed. Such medicines are likewise known as *drugs* ; and the department treating of them under the name of Pharmacographia (from φάρμακον, *a medicine*, and γράφω, *I write*).

Pharmacy (from φάρμακον, *a medicine*) treats of the collection, preparation, and preservation of medicines, and in its extended sense also includes all that relates to their dispensation.

Therapeutics (from θεραπεύω, *I cure*) comprises all that has reference to the actions and uses of medicines. This department has also been termed Pharmacodynamics (from φάρμακον, *a medicine*, and δύναμις, *power*).

Pharmacology, again, is either General or Special : the first department, as its name implies, treating generally of the actions, uses, and classification of medicines, &c., and the latter of medicines individually.

The above definitions of Materia Medica and its departments are essentially those of Pereira ; but Brunton, in his recent important work on ' Pharmacology, Therapeutics, and Materia Medica,' has adopted a somewhat different arrangement, and has used some of the terms in a different sense. Thus, after defining Materia Medica, as ' a knowledge of the remedies employed in medicine,' he subdivides it into Materia Medica proper, Pharmacy, Pharmacology, and Therapeutics. The two first departments essentially correspond to Pharmacognosy and Pharmacy of our arrangement ; but by Pharmacology, he understands ' a knowledge of the mode of action of drugs upon the body generally, and upon its various parts ;' and by Therapeutics, ' a knowledge of the uses of medicines in disease.'

The substances used as medicines are derived from the mineral, vegetable, and animal kingdoms ; those from the first are therefore termed *Inorganic*, and those from the two latter *Organic*. It is the object of the present work to describe the Special Pharmacology of Organic Substances, under the general name of Organic Materia Medica.

In treating of these organic substances we shall direct our attention more especially to those characters by which they may be recognised and their purity determined, or, in other words, to their *pharmacognosy.*

All botanical and zoological descriptions, and all details as to their chemistry, pharmacy, and therapeutics, must be obtained from special treatises on Botany, Zoology, Chemistry, Pharmacy, and Therapeutics ; or from more advanced works on Materia Medica.

Further, in using the term Organic Materia Medica, the author is fully aware that a distinction can no longer be strictly drawn between the Organic and Inorganic Materia Medica, but he employs the term in the conventional and ordinary sense, and as commonly employed by teachers and writers on Materia Medica. Again, all Products of Decomposition, as those of Fermentation and Destructive Distillation, are omitted, as their description properly belongs to chemistry.

With these exceptions, we shall describe as articles of the Organic Materia Medica, all the medicinal substances derived from plants and animals which are official in the British Pharmacopœia, as well as others of importance which are employed in the treatment of disease in this country, or kept generally by pharmacists. They will be treated of under the two heads of Vegetable Drugs and Animal Drugs, and will be arranged in the order of the natural historical relations of the organised beings which yield or produce them.

PART I.

VEGETABLE DRUGS.

UNDER the head of Vegetable Drugs we include all those which are derived from the Vegetable Kingdom, and which consist of plants, parts of plants, and their products and educts, except, as previously mentioned, the products of decomposition, as those of fermentation and destructive distillation. They will be described in the order in which the plants yielding them are arranged in the fifth edition of my ' Manual of Botany,' an arrangement which is founded, so far as the Phanerogamia are concerned, upon the system of De Candolle as modified by Bentham and Hooker, and adopted in their great work, ' Genera Plantarum.' It is as follows :—

The Vegetable Kingdom is first divided into two sub-kingdoms, namely :—Phanerogamia or Flowering Plants ; and Cryptogamia or Flowerless Plants.

Sub-kingdom 1. *Phanerogamia* or *Flowering Plants.*— This includes plants which have evident flowers ; and which are reproduced by seeds containing an embryo with one or more cotyledons.

Sub-kingdom 2. *Cryptogamia* or *Flowerless Plants.*— This includes those plants which have no flowers ; and which are reproduced by minute bodies called spores, which have no embryo.

SUB-KINGDOM I. PHANEROGAMIA or FLOWERING PLANTS. These are divided as follows :—

DIVISION I. *Angiospermia*, in which the ovules are dis-

tinctly enclosed in an ovary ; and are fertilised indirectly by the action of the pollen
on the stigma. Endosperm formed after fertilisation. It is divided thus :—

CLASS I. DICOTYLEDONES, in which the embryo is dicoty-
ledonous ; the germination exorhizal ; the stem
exogenous ; the leaves with a reticulated venation ; and the flowers commonly with a quinary
or quaternary arrangement. In this class we
have three sub-classes.

SUB-CLASS 1. POLYPETALÆ, with usually bisexual flowers,
which are commonly furnished with a
calyx and corolla, and the latter composed of distinct petals. This is divided
into three series as follows :—

Series 1. *Thalamifloræ*, that is, plants, the flowers of
which have usually the calyx, corolla,
and stamens distinct from one another ;
ovary superior ; and the stamens hypogynous.

Series 2. *Discifloræ.*—Thalamus furnished with a
disk, which is hypogynous or adnate to
the calyx or ovary, or bearing a series
of glands ; petals free ; stamens arising
from the disk and either hypogynous or
perigynous ; ovary superior, placentation
usually axile.

Series 3. *Calycifloræ.*—Calyx usually gamosepalous;
petals arising from the calyx or from a
perigynous disk ; stamens perigynous or
epigynous ; ovary superior or inferior.

SUB-CLASS 2. GAMOPETALÆ or COROLLIFLORÆ, with usually bisexual flowers ; calyx commonly
gamosepalous ; corolla gamopetalous
stamens inserted on the corolla or ovary,

or rarely separate from the corolla, and arising directly from the thalamus; ovary superior or inferior. Of this sub-class we have three series, as follows :—

Series 1. *Inferæ* or *Epigynæ*, in which the calyx is adherent and the ovary consequently inferior; stamens epigynous.

Series 2. *Superæ*, in which the calyx is inferior; the stamens inserted on the corolla, or rarely on the thalamus; ovary superior (except in *Vacciniaceæ*), and usually more than 2-celled.

Series 3. *Dicarpiæ* or *Bicarpellatæ*, in which the ovary is usually superior, and composed of two carpels, or rarely 1-3; stamens inserted on the corolla.

Sub-class 3. Monochlamydeæ or Incompletæ. — Flowers either have a calyx only (monochlamydeous), or without both calyx and corolla (achlamydeous); often unisexual. Of this sub-class we have two series, thus :—

Series 1. *Superæ*, in which the ovary is superior.

Series 2. *Inferæ* or *Epigynæ*, in which the ovary is inferior.

Class II. Monocotyledones, in which the embryo is monocotyledonous; the germination endorhizal; the stem endogenous; the leaves usually with a parallel venation; and the flowers with a ternary arrangement. This class may be divided into two sub-classes as follows :—

Sub-class 1. Petaloideæ.—Leaves with a parallel venation, or rarely reticulated, permanent or occasionally deciduous; floral envelopes (perianth) verticillate and usually coloured, rarely green or scaly, and sometimes absent. This sub-class may be divided into two series :—

Series 1. *Inferæ* or *Epigynæ*, in which the ovary is inferior, or rarely superior; and the perianth usually in two whorls, and both coloured.

Series 2. *Superæ*, in which the ovary is superior. Of this we have two sub-series.

Sub-series 1. *Apocarpæ*, in which the gynœcium is usually apocarpous, or rarely of one carpel.

Sub-series 2. *Syncarpæ*, where the gynœcium is syncarpous, or in some Palms apocarpous.

SUB CLASS 2. GLUMACE.E.—Leaves parallel-veined, permanent ; flowers glumaceous, that is, having no proper perianth, but imbricated bracts instead.

DIVISION II. *Gymnospermia*, in which the ovules are naked or not enclosed in an ovary, and fertilised directly by the action of the pollen. Endosperm formed before fertilisation.

SUB-KINGDOM II. CRYPTOGAMIA or FLOWERLESS PLANTS are those which have no proper flowers, that is, having no floral envelopes, stamens, or carpels, and which are reproduced by minute bodies termed spores, which have no embryo. This may be divided as follows :—

DIVISION I. *Cormophyta.*—Plants with commonly roots, stems, and leaves, and with vascular tissue ; or the latter is imperfect or entirely absent.

DIVISION II. *Thallophyta.*—Plants without any distinction of roots, stems, and leaves, and which are entirely composed of parenchymatous tissue.

The following is a tabular arrangement of the above system :—

SUB-KINGDOM I. PHANEROGAMIA or FLOWERING PLANTS.

DIVISION I. ANGIOSPERMIA.

  CLASS I. DICOTYLEDONES.

    SUB-CLASS 1. POLYPETALÆ.

      Series 1. Thalamifloræ.

        2. Discifloræ.

        3. Calycifloræ.

    SUB-CLASS 2. GAMOPETALÆ or COROLLIFLORÆ.

      Series 1. Inferæ or Epigynæ.

        2. Superæ.

        3. Dicarpiæ or Bicarpellatæ.

    SUB-CLASS 3. MONOCHLAMYDEÆ or INCOMPLETÆ.

      . Series 1. Superæ.

        2. Inferæ or Epigynæ.

  CLASS II. MONOCOTYLEDONES.

    SUB-CLASS 1. PETALOIDEÆ.

      Series 1. Inferæ or Epigynæ.

        2. Superæ.

          1. Apocarpæ.

          2. Syncarpæ.

    SUB-CLASS 2. GLUMACEÆ.

DIVISION II. GYMNOSPERMIA.

SUB-KINGDOM II. CRYPTOGAMIA or FLOWERLESS PLANTS.

DIVISION I. CORMOPHYTA.

DIVISION II. THALLOPHYTA.

Sub-kingdom I.

PHANEROGAMIA OR FLOWERING PLANTS.

Division I. *ANGIOSPERMIA.*

Class I. DICOTYLEDONES.

Sub-class I. POLYPETALÆ.

Series I. *THALAMIFLORÆ.*

Order 1.—RANUNCULACEÆ.

1. HELLEBORUS NIGER, *Linn.*
Black Hellebore. Christmas Rose.

(Bentley and Trimen's 'Medicinal Plants,' vol. i. plate 2.)

Habitat.—Central and Southern Europe. It is a favourite plant in our gardens from flowering in midwinter.

Part Used and Name.—HELLEBORI NIGRI RHIZOMA :—the dried rhizome and rootlets.

(*Not Official.*)

Hellebori Nigri Rhizoma.
Black Hellebore Rhizome.

Commerce.—Usually imported from Germany.

General Characters.—In commerce it is commonly known as *black hellebore root*, and either consists of the rhizome with the attached roots or rootlets, or usually the latter are more or less broken off and mixed with the rhizome. The rhizome varies in length from about one to three inches, and in thickness from one quarter to half an inch ; it is brownish-black in colour, very irregular in appearance, twisted, branched, and knotted, and marked on the upper surface with transverse ridges and slight longitudinal furrows. The rootlets are numerous, commonly several inches long, and about one-tenth of an inch in diameter, unbranched, sub-cylindrical, externally brownish-black, internally whitish, and presenting on a transverse section or fracture a thick bark surrounding a woody axis, which is undivided or indistinctly radiate. The odour of both the rhizome and rootlets is feeble, but has been compared to that of Senega Root ; and the taste is bitterish and somewhat acrid.

Principal Constituents.—Black hellebore rhizome contains two crystalline active principles, *helleborin* and *helleborein.* Both are glucosides, and are stated to be poisonous, and helleborin is also said to be highly narcotic. There is *no tannic acid* in black hellebore rhizome.

Adulterations and Substitutions.—In Germany both the rhizomes of *Helleborus niger* and of *Helleborus viridis* are in use, and we know of no very definite characters by which they may be distinguished ; but as the latter is the more expensive drug, it is not likely to be substituted for, or intermixed with, black hellebore rhizome except by accident ; and, moreover, as both would appear to have the same properties, the substitution of one for the other is of little importance. The substitution, however, of the rhizome of *Actæa spicata*, Linn., Baneberry, for that of black hellebore, which has been noticed on the Continent and in the United States, and also frequently by the author in this country, is an important one, as the two drugs have very different pro-perties. Baneberry may be readily distinguished by making a transverse section of one of its rootlets, when the woody axis will be seen to present a distinctly triangular, cruciate, or stellate arrangement of its component wedge-shaped bundles, according to its thickness. Moreover, as black hellebore rhizome contains no tannic acid, its infusion is not blackened by the addition to it of a solution of a persalt of iron ; whereas the latter immediately blackens an infusion of baneberry rhizome, from the presence in it of tannic acid.

Medicinal Properties.—Black hellebore rhizome is a drastic purgative, and is also regarded as emmenagogue and anthelmintic ; and in large doses it acts as an acro-narcotic poison. It is now but rarely used in this country, except for domestic animals.

---

2. DELPHINIUM STAPHISAGRIA, *Linn.*
Stavesacre.

(Bentley and Trimen's ' Medicinal Plants,' vol. i. plate 4.)

Habitat.—Countries almost throughout the Mediterranean region, from Portugal and Spain to Greece and Crete. It is also found in Asia Minor and the Canary Islands.

Official Part and Name.—STAPHISAGRIÆ SEMINA :—the dried ripe seeds.

Staphisagriæ Semina.
Stavesacre Seeds.

Commerce.—They are imported from Nîmes and other parts of the South of France, and also from Trieste.

General Characters.—Stavesacre seeds are irregularly triangular or obscurely quadrangular in form (*fig.* 1, *a, a*), arched on one side, and each weighing, on an average, a little over half a grain. The testa has a blackish-brown colour in the fresh seeds, but it becomes dull greyish-brown

a   *a*   *b*

FIG. 1.—*Stavesacre Seeds* (magnified).—*a, a.* Entire seeds. *b.* Vertical section of a seed.

in those that have been long kept ; its surface is deeply pitted and wrinkled. The nucleus is soft and oily ; it consists essentially of a whitish albumen, with a small straight embryo at one end (*fig.* 1, *b*). There is no very perceptible odour ; but the taste is bitter and nauseous, followed ultimately, after chewing, by tingling and burning.

Principal Constituents.—These seeds are said to contain no fewer than four alkaloids, the principal of which is *delphinine*, which is chiefly or entirely confined to the testa. An acid called *delphinic acid* has likewise been found, and a *fatty oil.* The *fatty oil* belongs to the non-drying class of oils, and is contained in the proportion of about 26 per cent.

Medicinal Properties.—These seeds are emetic, purgative, and anthelmintic, but are no longer used internally on account of their violent action. They are also narcotic, and in large doses poisonous. They are chiefly used externally as a parasiticide. The alkaloid delphinine has also been used externally in neuralgia, etc., like aconitine.

Official Preparation.—Unguentum Staphisagriæ.

---

3. ACONITUM NAPELLUS, *Linn.*
Monkshood. Aconite.

(Bentley and Trimen's 'Medicinal Plants,' vol. i. plate 6.)

Habitat.—Widely distributed in the mountainous districts of Europe, Asia, and North America ; and has become naturalised in a few places in the West of England and in South Wales.

Official Parts and Names. — 1. ACONITI FOLIA :—the fresh leaves and flowering tops, gathered when about one-third of the flowers are expanded, from plants cultivated in Britain. 2. ACONITI RADIX :—the root, collected in the winter or early spring before the leaves have appeared, from plants cultivated in Britain, and carefully dried ; or imported in a dried state from Germany.

1. Aconiti Folia.
Aconite Leaves.

Collection.—The fresh leaves are directed to be used, because their properties, although not very sensibly injured by careful drying, are in all cases much weakened by keeping. Hence in those countries where they cannot be obtained in a fresh state, the recently dried leaves should alone be employed, or, still better, aconite root should be exclusively used.

The fresh leaves and flowering tops are also directed to be gathered when about one-third of the flowers are expanded, for reasons which have been explained by the author (*Pharm. Journ.* 2nd ser. vol. iii. page 475) as follows :— There are two series of organic products formed in plants, those of one series being especially concerned in their growth and development, and those of the other series playing no active part in plants after their production, and being also commonly formed later in their life. In the process of flowering the only products that are taken up are those which are concerned in the growth and development of new tissues ; while the other series, by the removal of these nutritive products, become more concentrated, and the organs in which they are formed, by being left for a longer period, have also time to elaborate them more completely. Hence, in such cases, the preparations obtained from them are not only more active, but also more stable. When herbaceous plants or their parts are therefore ordered in the British Pharmacopœia, they are always directed to be taken when the flowering stage has advanced to some extent, but the exact degree has been found to vary in different plants, and is thus indicated in each case under the head of Collection. In all cases, however, care should be taken that the flowering stage has not so far advanced as to weaken the active vitality of the nutritive organs.

General Characters.—Leaves alternate, with long furrowed petioles, very deeply cut palmately into five or three segments, which are again deeply and irregularly.divided into oblong, acute, narrow, somewhat wedge-shaped lobes, dark-green, smooth, and shining above, paler below ; exciting slowly, when chewed, a sensation of tingling and numbness ;

no marked odour. Flowers large, irregular, dark blue, and arranged in a somewhat loose terminal raceme.

Principal Constituents.—Aconitine in small proportion (*see* Aconiti Radix), and *aconitic acid.*

Medicinal Properties and Official Preparations.—(See Aconiti Radix).

2. Aconiti Radix.

Aconite Root.

Collection and Commerce.—The roots, or tubers as they are frequently termed, are directed to be collected in the winter or early spring before the leaves have appeared, because, like all other roots and underground stems, they are most active when the herbaceous parts are dead, or before new structures are developed, as at such times their active constituents are concentrated in them, and exist in their most perfectly formed and stable condition.

They are principally imported from Germany, as aconite is but little cultivated in Britain. This is much to be regretted, as the imported roots being obtained from wild plants cannot be collected with any certainty at the proper period, and are commonly of a mixed character and of inferior quality.

General Characters.—In a *fresh* state, when obtained in the winter or early spring, aconite or monkshood root is from three to six inches, or rarely more, in length, and from three-quarters of an inch to an inch or more in thickness (*fig.* 2, *a*, *b*). It is distinctly conical in form, and tapers downwards to a threadlike point, with numerous branched rootlets on its sides. If obtained in the summer, a second root, or, rarely, a third, is attached to its upper extremity by a short branch. Both the root and rootlets are coffee-coloured or dark brown externally, and white internally. There is no very marked odour; and the taste is at first only very slightly bitter, but in a few minutes a peculiar sensation of tingling and numbness is produced.

Notwithstanding the above marked characters of aconite

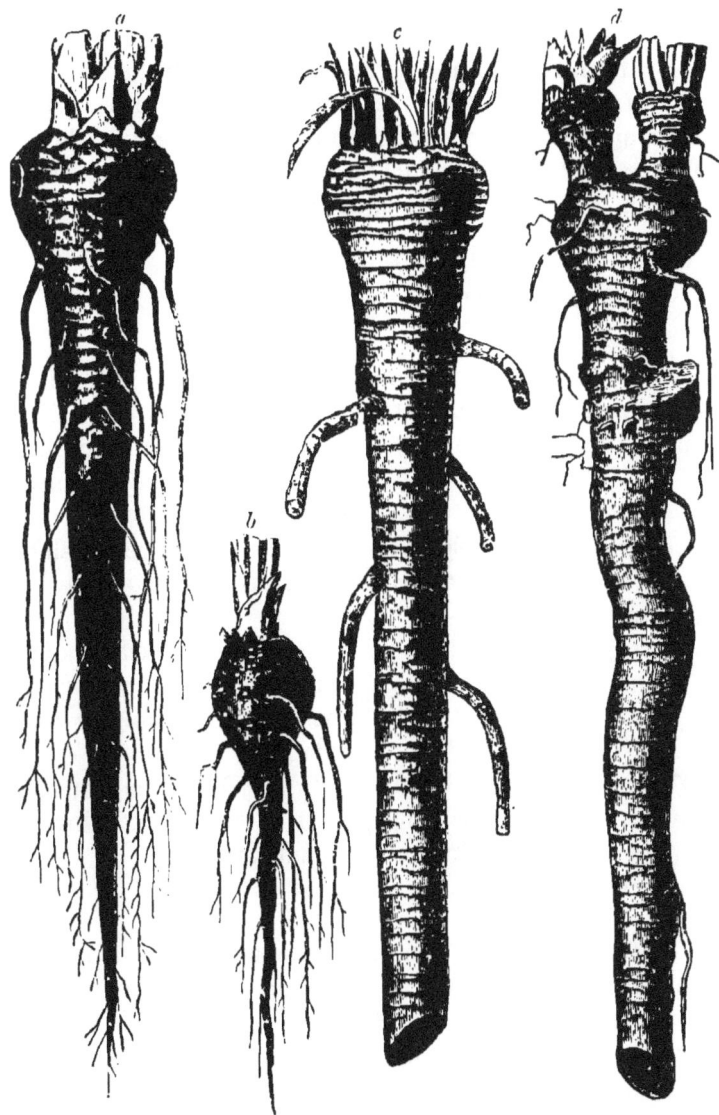

FIG 2.—*a, b. Fresh Aconite Roots. c, d. Fresh Horseradish Roots.*

root, several fatal cases of poisoning have occurred at those periods of the year when the leaves are absent, from its having been scraped and served up at table as horseradish root (*Armoraciæ Radix*); and in order to guard against such a highly dangerous substitution, the author published a paper in the 'Pharmaceutical Journal' as far back as 1856, in which the distinctive characters of the two roots were tabulated as follows :—

Aconite Root.	Horseradish Root.
(Fig. 2, *a, b.*)	(Fig. 2, *c, d.*)
Form.—Conical, and tapering perceptibly and rapidly to a fine point.	*Form.*—Slightly conical at its base or upper end, then cylindrical, or nearly so, and almost of the same thickness for several inches.
Colour. –Coffee-coloured, or more or less distinctly earthy-brown externally.	*Colour.*—Pale yellowish or brownish-white externally.
Odour.—Merely earthy, or somewhat radish-like.	*Odour.* — Especially developed upon being scraped, when it is very pungent and irritating.
Taste.—At first slightly bitter, but afterwards producing a peculiar tingling and numbness.	*Taste.*—Very pungent, and bitter or sweet according to circumstances.

To these distinctive characters we may add that a pinkish colour is soon developed in the scrapings of aconite root when they are exposed to the air ; but no such change of colour occurs under similar circumstances with those of horseradish root. It is possible that the common practice of putting into the ground the upper portion of horseradish root, after the lower part has been scraped away and used, may have led to these two roots being confounded, for, as seen by our description, it is here only that there is any resemblance in form between them.

In its *dried state*, in which alone it is official, aconite root is commonly from about two to three inches long, and from a half to about three-quarters of an inch at its upper extremity, where it is usually crowned with the remains of the stem, or there is a corresponding scar.   It is more or less conical in form, usually much shrivelled longitudinally, and marked to a varying extent with the scars or bases of broken rootlets.   The colour externally is dark brown, and internally whitish; it breaks with a short fracture, the surface presenting a somewhat mealy character; and if not hollow, which it is at times, there is an irregularly stellate central cellular axis or pith, with from six to eight rays, commonly seven (*fig.* 3).   It has no marked odour ; but its taste is somewhat bitterish-sweet at first, but exciting slowly, when chewed, a peculiar sensation of tingling and numbness, which lasts for some time.

FIG. 3.——Transverse section of *Aconite root.*

Substitutions and Adulterations.—As already noticed under the head of ' Collection,' the imported root is commonly of very inferior quality.   The roots of other species of *Aconitum* may be also not unfrequently found mixed with them, and also at times the roots of other plants.   Thus Holmes has found the aromatic roots of *Imperatoria Ostruthium* (*Pharm. Journ.* 3rd ser. vol. vii. p. 749).

Principal Constituents.—*Aconitic acid*, and at least five alkaloids, which together constitute only about 0·07 per cent.   These alkaloids are, *aconitine, pseudaconitine, aconine, pseudaconine*, and *picraconitine* ; commercial aconitine being a mixture in varying proportions of these basic substances.

Medicinal Properties.—Aconite leaves and aconite root are sedative, anodyne, diuretic, and antiphlogistic, and in improper doses virulently poisonous.   The liniment, when locally applied to a painful part, produces at first a sensation

c

of tingling, which is soon followed by numbness and cessation of pain. Aconite root is by far the most active part of the plant ; it is said to have six times the strength of the leaves. Aconitine has similar but far more powerful properties.

Official Preparations.

1. Of ACONITI FOLIA :—

    Extractum Aconiti. *Dose.*—$\frac{1}{4}$ to 1 grain.

2. Of ACONITI RADIX:—

    Aconitina, which is also used in the preparation of Unguentum Aconitinæ.

    Linimentum Aconiti.

    Tinctura Aconiti. *Dose.*—5 to 15 minims.

OTHER ACONITE ROOTS.

Besides the official root, some other aconite roots possess similar properties, of which the following is the most important :—

Aconitum ferox, Wallich, *Nepal* or *Indian Aconite* (Bentley and Trimen's ' Medicinal Plants,' vol. i. plate 5). This kind of aconite root, as seen in commerce in a dried state, varies in length from two to four inches, and from a half to nearly two inches in diameter at its base. It is conical when entire, much shrivelled longitudinally, and marked here and there with the scars of broken-off rootlets. In colour externally, the roots are blackish-brown, except at their projecting portions, where from friction they are frequently whitish ; internally, they are commonly brown, very hard, horny, and translucent, but in some cases mealy and white. They have no odour, but a similar taste to the official aconite root, but more intense.

The roots are imported direct from India, occasionally in large quantities, but the supply is irregular.

They have a similar composition in all essential particulars to the roots of *Aconitum Napellus*, but *pseudaconitine* is the principal alkaloid contained in them.

4. CIMICIFUGA RACEMOSA, *Elliott.*

Black Snakeroot.

Synonym.—ACTÆA RACEMOSA, *Linn.*

(Bentley and Trimen's ' Medicinal Plants,' vol. i. plate 8.)

Habitat.—United States of America as far south as Florida ; it is also found in Canada.

Official Part and Name.—CIMICIFUGÆ RHIZOMA :—the dried rhizome and rootlets.

Cimicifugæ Rhizoma.

Cimicifuga.

Synonym.—Actææ Radix.

Commerce.—It is imported from the United States of America.

General Characters.—As seen in commerce, cimicifuga, or Black Snakeroot as it is also commonly called, consists of the rhizome, or portions of the same, with a variable number of rootlets arising from it below, or the latter are detached from, and mixed with, the rhizomes. The rhizome is two or more inches long, and half an inch or more thick ; flattened-cylindrical in form, hard, branched, marked above with the remains of former aerial stems and the scars of fallen leaves, and giving off below numerous small wiry brittle branched rootlets, which vary in length, and in thickness averaging about one-twelfth of an inch. Both rhizome and rootlets are dark brown or blackish, have a bitter and somewhat acrid taste, but are almost inodorous, except when bruised and moistened with warm water, when they have a slight narcotic odour. The rhizome and rootlets have a close fracture, the exposed surface of the former exhibiting a large central whitish pith surrounded by a variable number of stellately arranged woody bundles (*fig.* 4, *a*), surrounded by a thick bark. A transverse section of a rootlet shows the ligneous portion with commonly four woody bundles arranged in a cross-like manner (*fig.* 4, *b*), or in the smaller rootlets the woody bundles are triangularly arranged (*fig.* 4, *b*), and in the larger rootlets in a radiate or stellate manner.   In these respects the arrangement of the woody wedges resembles that of the rootlets of *Actæa spicata*, which we have previously mentioned as frequently substituted for black hellebore (see page 10).

Principal Constituents. — A neutral crystalline principle, the composition of which has not been determined, but it forms very acrid solutions with rectified spirit and chloroform ; *resin* ; and *tannic acid.* The substance employed in the United States by the eclectic practitioners, called *cimicifugin* or *macrotin*, and which is found in the form of a dark-brown powder or in scales in the proportion of nearly 4 per cent., is an impure resin which is precipitated from a concentrated tincture of cimicifuga by the addition of water.   The active properties of cimicifuga, so far as they have been determined, are due to a large extent at least to this so-called cimicifugin.   An infusion is blackened by a persalt of iron.

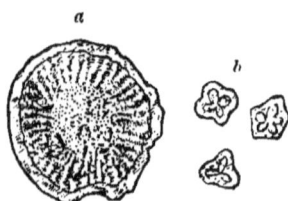

FIG. 4. — Transverse sections of *Cimicifuga. a.* Rhizome. *b.* Rootlets. (After Maisch.)

Medicinal Properties. — Stomachic, cardiac tonic, and expectorant ; it is regarded as a valuable remedy in acute rheumatism, chorea, and bronchitis.   A concentrated tincture has been recommended in the United States as a valuable external application for the purpose of reducing inflammation.

Official Preparations.

Extractum Cimicifugæ Liquidum.   *Dose.*—3 to 30 minims.

Tinctura Cimicifugæ.   *Dose.*—15 to 60 minims.

ORDER 2.—MAGNOLIACEÆ.

ILLICIUM ANISATUM, *Linn.*
Star-Anise.

(Bentley and Trimen's ' Medicinal Plants,' vol. i. plate 10.)

Habitat.—South-Western China. Said to have been introduced at an early period into Japan.

Official Parts or Products and Names.—1. ANISI STELLATI FRUCTUS : the dried fruit. From plants cultivated in China. 2. OLEUM ANISI :—the oil distilled in Europe from anise fruit (*see* Pimpinella Anisum, *Linn.*) ; or in China from star-anise fruit (Illicium anisatum, *Linn.*).

1. Anisi Stellati Fructus.
Star-Anise Fruit.

This is only introduced into the British Pharmacopœia from being one of the sources of the official Oleum Anisi.

General Characters.—Star-Anise fruit is usually com-

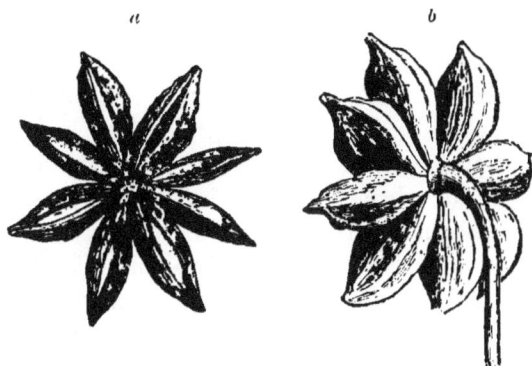

FIG. 5.—*Star-Anise Fruit.* *a.* Upper surface. *b.* Lower surface with stalk attached.

posed of eight fully developed carpels, diverging horizontally in a stellate manner from a central column, which is commonly placed on a short more or less oblique stalk ;

hence the name star-anise applied to it (*fig.* 5, *a, b*). This fruit is more properly designated as a kind of capsule, as its constituent carpels are united at the base, but commonly the component parts are regarded as so many follicles. Each carpel or follicle is half an inch or more long, boat-shaped terminated at the apex by a short nearly straight beak irregularly wrinkled, somewhat woody, of a rusty-brown colour, and usually open on its upper (ventral) margin, so as to expose a smooth reddish-brown cavity, and a solitary flattish-oval polished shining brownish-red-yellow seed. Odour and taste of both pericarp and seed closely resembling anise fruit ; but the seeds are somewhat less aromatic than the pericarp, and have a more oily taste.

Principal Constituents.—The chief constituent is the official volatile oil, which is described below. The oil is most abundant in the pericarp. Star-Anise fruit is also rich in sugar, probably cane sugar (*sucrose*), and the seeds contain much fixed oil.

Substitution.—The fruit of a variety of *Illicium anisatum*, or, according to some, a distinct species, *Illicium religiosum*, Siebold, derived from Japan, and known as *Sikimi* or *Skimmi Fruit*—has been substituted for the official fruit. It possesses poisonous properties, and hence its detection is most important. The two fruits have, as a rule, the same number of carpels, and these are similarly arranged, so that they have a close resemblance ; but the carpels of *skimmi fruit* are shrivelled as if immature, more woody, have a more acute beak, which is usually distinctly turned upwards, an unpleasant taste, and no anise-like odour, but one which has been described by some as clove-like, and by others as resembling bay leaves, nutmeg, &c. Its poisonous principle has been called *sikimine*.

2. Oleum Anisi.

Oil of Anise.

The official sources of this oil, as already stated, are as follows :—

The oil distilled in Europe from anise fruit ; or, in China, from star-anise fruit. The latter oil is sometimes known as *Oleum Badiani*, and may be distinguished as *Oil of Star-Anise.*

General Characters.—The amount of volatile oil obtainable from star-anise fruit has been variously estimated, but usually it averages from 4 to 5 per cent. It is colourless or very pale yellow; with the odour of the fruit, and an aromatic sweetish taste. Its specific gravity, which increases with age, varies from 0·977 to 0·985. It is essentially composed of *anethol* or *anise camphor*, and is entirely soluble in alcohol or ether ; its rotatory power is feebly levogyre.

The oils obtained from the two fruits are identical in composition, and are nearly the same in most of their characters, although from a slight difference in taste and odour they can be distinguished by dealers. But they congeal at different temperatures; for while the oil from anise fruit congeals at temperatures between 50° and 60° F. (10° to 15°·5 C.), and may remain solid at 62° or 63° F. (16°·7 to 17°·2 C.); that from star-anise fruit congeals at about 35° F.

Substitution and Adulterations.—The volatile oil of skimmi fruit, already described as having been substituted for true star-anise fruit, has been also sometimes imported as star-anise oil. But the absence of any anise-like odour ought at once to lead to its detection. Its specific gravity is also higher, being 1·006, and it does not congeal even when cooled to 4° below zero of Fahrenheit. The poisonous principle of skimmi fruit or Japanese star-anise fruit, and provisionally named *sikimine*, is not present in the volatile oil.

Oil of anise is frequently adulterated with spermaceti, wax, or camphor. The two former may be readily distinguished by their insolubility in cold alcohol, the latter by its odour.

Medicinal Properties.—Star-Anise fruit is not used medicinally or otherwise in this country; but the oil is employed for its aromatic, carminative, and stimulant properties.

Official Preparations.

1. Of ANISI STELLATI FRUCTUS :—
   Oleum Anisi. *Dose.*—1 to 4 minims.
2. Of OLEUM ANISI :—
   Essentia Anisi. *Dose.*—10 to 20 minims.
   Tinctura Camphoræ Composita. *Dose.*—15 minims to 1 fluid drachm.
   Tinctura Opii Ammoniata. *Dose.*—½ to 1 fl. drachm.

ORDER 3.—MENISPERMACEÆ.

1. CHONDRODENDRON TOMENTOSUM, *Ruiz & Pavon.*
Pareira Brava.

(Bentley and Trimen's ' Medicinal Plants,' vol. i. plate 11.)

Habitat.—Near Rio Janeiro, and in other parts of Brazil, and also in Peru.

Official Part and Name.—PAREIRÆ RADIX :—the dried root.

Pareiræ Radix.
Pareira Root.

Commerce.—It is imported from Brazil.

General Characters.—In long, nearly cylindrical, branched or unbranched, more or less twisted pieces (*fig.* 6, *a*), varying in thickness from about three-quarters of an inch to two or more inches; covered with a thin, somewhat loosely attached, blackish-brown bark; and marked externally with deep irregular longitudinal furrows, and numerous transverse ridges and furrows.   Internally it has a yellowish or brown-ish-grey colour; the wood being porous and arranged in well-marked concentric or more or less eccentric circles, which are sepa-rated into wedge-shaped portions by large medullary rays (*fig.* 6, *b*, *c*) ; each zone of wood being also separated from its neighbour by a wavy ring of medullary substance, which is pale at first, but becomes darker by keeping and exposure to the air.   It breaks with a coarse fibrous fracture, but when cut or shaved it has a waxy character. It is inodorous, but with a bitter taste.

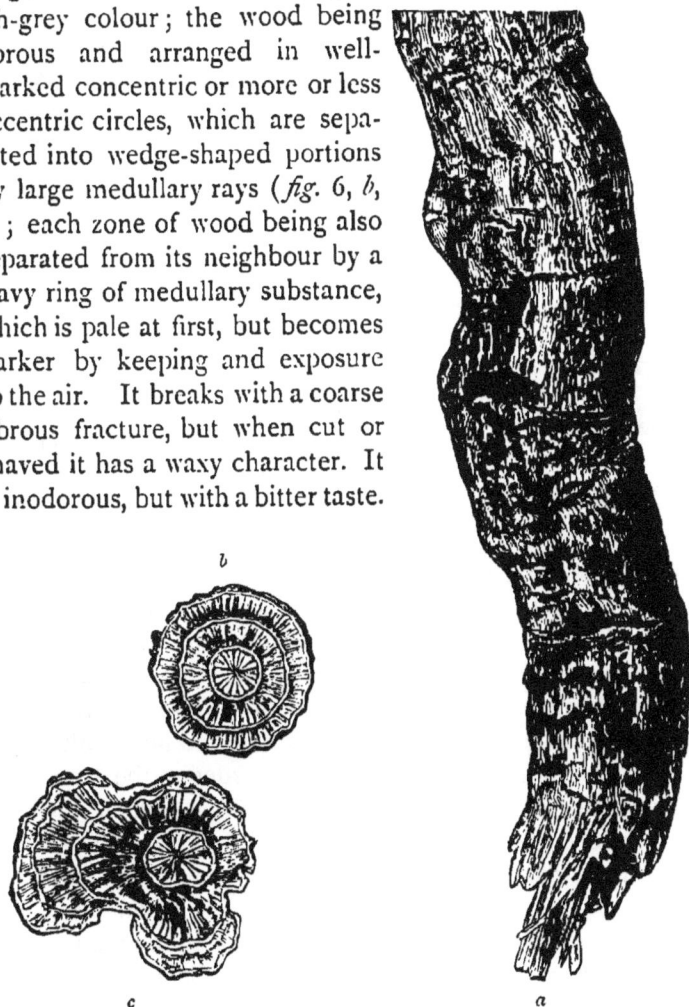

FIG. 6.—*Pareira Root.* *a.* Portion of a root. *b, c.* Transverse sections of roots. (After Hanbury.)

Principal Constituents.—A supposed peculiar crystalline principle, which has been named *pelosine* or *cissampeline*, is usually designated as the active bitter constituent of pareira root ; but Flückiger has shown this alkaloid to be identical with *buxine* from the bark of *Buxus sempervirens*, Linn., and *beberine* from the bark of *Nectandra Rodiæi*, Schom. Pareira root also contains much *starch* ; hence its decoction, when cold, is turned inky bluish-black by solution of iodine.

Substitutions.—The stems of *Chondrodendron tomentosum* are sometimes imported and sold for the official root. They have the same bitterish taste as the root, but may be distinguished by having a small, although evident pith, which is absent in the root, and by their lighter coloured greyish bark, which is also commonly more or less covered by whitish lichens.

FIG. 7.—Transverse section of the stem of *Cissampelos Pareira*, Linn.

Several other roots and stems have also at various times been substituted for the official root, but the characters of the latter as given above are so well marked, that no difficulty ought to be experienced in detecting the false from the true drug. The more important of these substitutions are the stems and roots of *Cissampelos Pareira*, Linn. ; which is very different, both in external appearance, transverse section (*fig.* 7), and otherwise ; White Pareira Brava, from *Abuta rufescens*, Aublet ; Yellow Pareira Brava, from, it is supposed, *Abuta amara*, Aublet; and the Common False Pareira Brava, the exact botanical source of which is unknown, although certainly from a plant of the same order as the true root. This latter spurious root had at one time almost entirely superseded the original and true drug : but it is readily distinguished by its outer surface being of a lighter brown colour ; by cutting distinctly fibrous like wood generally, instead of like wax ; by its rings being commonly very eccentrically arranged (*fig.* 8), and of different structure ; and by the action of iodine, for whereas a cool decoction of the spurious drug is not perceptibly coloured by solution of iodine, that of the true root is turned by it of an inky bluish-black colour.

Medicinal Properties.—Pareira root possesses mild tonic and diuretic properties, and is stated to exercise an almost

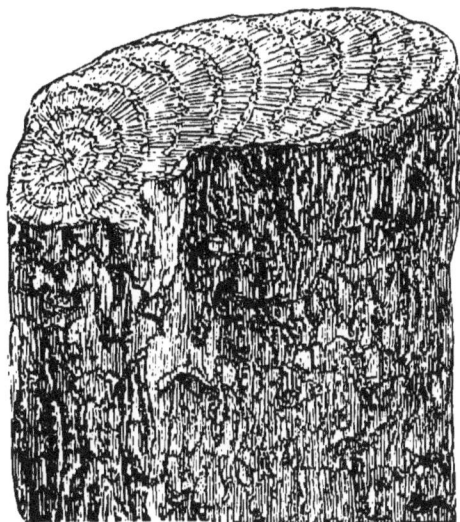

FIG. 8. Transverse section of *False Pareira Root.* From an undetermined Menispermaceous plant. (After Hanbury.)

specific influence over the mucous membrane of the genito-urinary organs.

Official Preparations.

Decoctum Pareiræ. | Extractum Pareiræ.

Extractum Pareiræ Liquidum.

2. JATEORHIZA CALUMBA, *Miers.*

Calumba. Columbo.

Synonym.—COCCULUS PALMATUS, *DC.*

(Bentley and Trimen's 'Medicinal Plants,' vol. i. plate 13.)

Habitat.—Eastern tropical Africa, and especially along the course of the Lower Zambesi.

Official Part and Name.—CALUMBÆ RADIX :—the dried transversely cut slices of the root.

Calumbæ Radix.

Calumba Root.

Collection and Commerce.—From the forests of Eastern Africa, between Ibo and the banks of the Zambesi. It is either shipped directly from Mozambique and Zanzibar, or it is obtained indirectly by way of Bombay and some other Indian ports.

General Characters.—In irregular, flattish, rounded or somewhat oval pieces, varying from one inch to two inches or more in diameter, and from one-eighth to half an inch or more in thickness. The cortical portion is thick, covered by a closely attached, wrinkled, brownish-yellow coat, and separated from the central woody portion by a fine dark-coloured line. The woody portion is more or less concave on both surfaces, from contraction whilst being dried ; of a greyish- or greenish-yellow colour, and marked by medullary rays and faint concentric circles. The pieces have a short mealy fracture, and are readily reduced to powder. They have a feeble musty odour, and a bitter mucilaginous taste.

Principal Constituents.—A peculiar crystalline colourless neutral bitter principle called *columbin* ; a bitter crystalline yellow alkaloid termed *berberine*, which exists in combination with a peculiar acid, *columbic acid*, and which, together with berberine, gives calumba root its yellow colour. To these three constituents, but more especially to columbin, the bitter taste and medicinal properties of calumba are due. *Starch* is also present to a large amount, constituting on an average about one-third of its weight ; hence a decoction, when cold, is coloured deep bluish-black by solution of iodine.

Adulterations and Substitutions.—So far as our experience goes, calumba root, although varying much in quality, is rarely or ever adulterated in this country. Moreover, the only substitution we have any direct knowledge of is the substance known as calumba wood, first described by

Hanbury, and afterwards traced by him to *Coscinium fenestra-tum*, Colebr., a native of Ceylon, and belonging, like the true calumba plant, to the order Menispermaceæ. It is found in the form of transverse slices of the stem (*fig.* 9), but readily distinguished by their difference of structure, greater smooth-ness and hardness, and from not being contracted in the centre. Of late years calumba wood has also been offered in the London drug sales as the stems of the true calumba plant ; but, as just noticed, its geographical, as well as botanical, source is different.

Fig. 9.—Transverse section of the stem of *Coscinium fenestratum*, Colebr., known as *Calumba Wood*. (After Hanbury.)

Medicinal Properties.—Calumba root is a pure bitter tonic and stomachic, and from containing neither tannic or gallic acid, it may be combined with salts of iron and alkalies.

Official Preparations.

Extractum Calumbæ.	Mistura Ferri Aromatica.
Infusum Calumbæ.	Tinctura Calumbæ.

---

3. ANAMIRTA PANICULATA, *Colebr.*

Cocculus Indicus.

(Bentley and Trimen's ' Medicinal Plants,' vol. i. plate 14.)

Habitat.—Eastern side of Indian Peninsula, Ceylon, and the Malayan Islands.

Part Used and Name.—COCCULUS INDICUS :—the dried fruit.

·(*Not Official.*)

Cocculus Indicus.

Cocculus Indicus.

Collection and Commerce.—It is now imported from Bombay and Madras ; but formerly it was known under the names of Levant Nut and Levant Shell, from being brought to Europe by way of the Levant.

General Characters.—Cocculus Indicus is reniform or reniform-ovoid, and usually less than half an inch in length, or about the size of a small hazel-nut. It is commonly termed a berry, but is more correctly described as a drupaceous fruit. It consists externally of a thin blackish-brown wrinkled skin, which is composed of the combined mesocarp and epicarp, within which is a thin whitish woody shell or endocarp. On the concave side of the fruit the shell projects deeply into the interior, and bears upon its surface a solitary yellowish-white oily reniform or semi-lunar seed. This seed never wholly fills the cavity of the fruit, as it contracts in drying, and by keeping it becomes more and more shrivelled and wasted, so that in old samples of Cocculus Indicus it is not uncommon to find the endocarp almost empty. Hence the test of its goodness was formerly, that ' the seed should fill at least two-thirds of the shell.' Cocculus Indicus is inodorous, and the endocarp almost tasteless, but the seed is very bitter.

The only drug likely to be confounded with Cocculus Indicus is that which is commonly known as Bay Berry, the fruit of *Laurus nobilis*, Linn.; but the latter is commonly larger, distinctly oval in form, has a thin brittle pericarp, and its solitary seed has an agreeable aromatic odour, firm consistence, oval form, is readily separable into two equal portions, and lies free in, and fills the cavity of, the fruit.

Principal Constituents.—The seed, the only part used, owes its active properties to the presence of a white crystalline, non-nitrogenised, intensely bitter, and very poisonous neutral principle called *picrotoxin.* The pericarp contains two tasteless crystallisable alkaloids of the same composition termed *menispermine* and *paramenispermine;* these are in combination with a peculiar acid, known as *hypopicrotoxic acid.*

Medicinal Properties.—Cocculus Indicus and picrotoxin are narcotic and poisonous. They are not used internally, but sometimes externally in the form of an ointment in certain skin diseases, and to destroy vermin.

Preparation.—An Unguentum Cocculi was official in the British Pharmacopœia of 1864, and directed to be prepared by mixing 80 grains of the seeds of Cocculus Indicus with 1 ounce of prepared lard. Or, in place of the seeds, 10 grains of picrotoxin may be used.

Order 4.—BERBERIDACEÆ.

PODOPHYLLUM PELTATUM, *Linn.*
May-Apple.

(Bentley and Trimen's 'Medicinal Plants,' vol. i. plate 17.)

Habitat.—United States and Canada.

Official Parts or Products and Names.—1. Podophylli Rhizoma :—the dried rhizome and rootlets. 2. Podophylli Resina :—the resin.

1. Podophylli Rhizoma.
Podophyllum Rhizome.

Synonym.—Podophylli Radix.

Commerce.—It is imported from the United States of America.

General Characters.—In commercial specimens the rootlets are either entirely absent, or when present they may be attached to the rhizomes, or in a separated state and mixed with them. The *rhizome* is horizontal, simple, or rarely branched, variable in length, usually two or more inches; and from about one-fifth to one-third of an inch thick. It is flattened-cylindrical, and when of any length it presents one or more large irregular tuberosities, which are marked above by a depressed circular scar, indicating the point where the aerial stem was attached, and giving off below a variable number of rootlets, or if these are broken up their position is indicated by whitish scars. Externally it is smooth, or somewhat furrowed or wrinkled, and of a dark reddish-brown or reddish-yellow colour ; it is brittle, and its fracture is short, whitish, and mealy. The *rootlets* are small, vary in length, resemble in colour but are somewhat paler than the rhizome, and very brittle. Both rhizome and rootlets have a faint narcotic odour, more especially when moistened with warm water ; and a bitterish acrid nauseous taste. The powder is of a greyish-yellow colour.

Principal Constituents.—The active constituent is the official *resin* (*see* Podophylli Resina) ; but the rhizome also contains *starch*, *sugar*, and one or more acid and basic principles which have no medicinal importance.

2. Podophylli Resina.

Resin of Podophyllum.

General Characters.—This resin, which is contained in the proportion of from $3\frac{1}{2}$ to 5 per cent. in the rhizome, and as prepared according to the directions of the British Pharmacopœia, is an amorphous powder, varying in colour from pale yellow to deep orange-brown ; inodorous, but with an acrid bitter taste. This resin is partly neutral, and partly acid (*podophyllinic acid*), both portions of which are soluble in rectified spirit, but only partially soluble in pure ether. The resin is also soluble in ammonia, from which solution it is precipitated by acids, and from its solution in rectified spirit by water.

Medicinal Properties.—Alterative, cathartic, cholagogue.

Both rhizome and resin are in use in the United States, but in Britain the resin is alone commonly employed.

Official Preparations.

1. Of PODOPHYLLI RHIZOMA :—
    Resina Podophylli.   *Dose.*—¼ to 1 grain.
2. Of PODOPHYLLI RESINA :—
    Tinctura Podophylli.   *Dose.*—15 minims to 1 fluid drachm.

ORDER 5.—PAPAVERACEÆ.

1. PAPAVER SOMNIFERUM, *Linn.*

Opium Poppy. White Poppy.

Bentley and Trimen's ' Medicinal Plants,' vol. i. plate 18.)

Habitat.—Not now known in a thoroughly wild condition, but probably originally a native of South-Eastern Europe and Asia Minor. In England it occurs sporadically and not unfrequently, especially in the south.

Official Parts or Products and Names.—1. PAPAVERIS CAPSULÆ:—the nearly ripe dried capsules. From plants cultivated in Britain. 2. OPIUM :—the juice obtained in Asia Minor by incision from the unripe capsules, inspissated by spontaneous evaporation.

Any ordinary variety of opium may be employed as a source of alkaloids, and of extract of opium of official strength ; but, when otherwise used for officially recognised purposes, opium must be that obtained in Asia Minor, and must be of such a strength that, when dried and powdered and the powder heated to 212° F. (100° C.) until it ceases to lose moisture, and the product tested by the appended method (*see* British Pharmacopœia), or any trustworthy method, it shall yield, as nearly as practicable, 10 per cent.

of morphine ; that is, 100 parts of such dry powdered opium shall yield not less than 9·5 parts, and not more than 10·5 parts, of morphine.

1. Papaveris Capsulæ.
Poppy Capsules.

Collection and Commerce.—Poppy capsules, or *Poppy heads* as they are commonly called, are directed to be collected when *nearly ripe*, because analysis proves that they then contain most morphine, and are consequently in their most active state ; but in practice this direction of the British Pharmacopœia is not very strictly adhered to. They are also ordered to be obtained from plants cultivated in Britain, and for this purpose the Poppy plant is grown to a limited extent in England, at Mitcham, Hitchin, near Banbury, and some other places. It is very largely cultivated in different parts of Europe, more especially in Germany, for its capsules and seeds. At Banbury the capsules are gathered about the end of August or commencement of September, and then dried in kilns, which process takes about twelve hours. The finest capsules are usually sold entire, while the smaller and less showy ones are broken up, divested of their seeds, and supplied to the pharmacist for making the preparations. It will be noticed, however, under the head of Official Preparations, that the seeds should be also used in making the Decoction of Poppy.

General Characters.—Poppy capsules are rounded, ovoid-rounded, or somewhat oblong in form, and vary in diameter from two to three inches, or from the size of a hen's egg to that of the fist. They are crowned above by the stellately-arranged stigmas, and are suddenly contracted below into a sort of neck, which is attached by an enlarged base to the cut end of the stalk. They have a yellowish or yellowish-brown colour externally, and are frequently dotted with blackish spots ; they are papyraceous in texture, brittle, and when cut transversely they are seen to present internally a variable number of thin, brittle, parietal placentas, directed towards the centre of the cavity, and a very large number of loose, small, reniform seeds, with a reticulated testa, and of a whitish, nearly black, or somewhat slate colour. The dark coloured seeds are commonly known as *maw seeds*, but this name is also given in some cases to poppy seeds of whatever colour. Neither the pericarp nor seeds have any odour ; but the pericarp is slightly bitter, and the seeds have an oily sweetish taste.

Principal Constituents.—Several chemists have found *morphine*, and generally in larger proportion in the nearly ripe capsules, and hence, as already noticed, they are then ordered to be collected. Traces of *meconic acid, narcotine,* and some other of the peculiar constituents of opium, have also been indicated, but their activity depends essentially on morphine. Two crystalline bodies, termed *papaverine* and *papaverosine*, have also been found in the pericarp.

The *seeds* contain none of the active constituents of opium, and are therefore entirely devoid of any of its narcotic effects. They yield by expression a *fixed oil*.

Medicinal Properties.—Similar, but much weaker and less trustworthy than opium when used internally. The decoction, when applied hot, forms a useful anodyne and demulcent fomentation.

Official Preparations.

Decoctum Papaveris.

Extractum Papaveris. *Dose.*—2 to 5 grains.

Syrupus Papaveris. *Dose.*—10 to 15 minims for an infant of three or four months old ; and for adults 1 fluid drachm.

In making Extract of Poppy and Syrup of Poppy the seeds are directed to be removed, and the pericarp alone used ; but no such direction is given with the Decoction of Poppy, the bruised poppy capsules being there alone mentioned, as the seeds contribute by their oily properties to the emollient quality of the decoction.

2. Opium.

Production, Extraction, and Collection.—The official opium is chiefly produced in the North-West of Asia Minor, but also in the north, and to some extent in other districts of the same country.

The mode of extracting opium is essentially the same in all countries where it is obtained, and consists in making transverse (Asia Minor and Egypt), or longitudinal (India) incisions into the unripe capsules, and subsequently collecting the exuded juice when concreted, but still in a soft state. As the opium produced in Asia Minor is specially mentioned in the pharmacopœia, it will be necessary for us to give generally the mode in which it is there obtained. This is as follows :—About the end of May, the plant (*Papaver somniferum*, var. *glabrum*, Boissier) arrives at maturity, and a few days after the petals have fallen, and when the capsule is of a light green hue, it is ready for incising, which operation is performed in the afternoon of the day, by making a transverse cut with a one-bladed knife about its middle, the incision being usually carried round until it nearly reaches again the point whence it commenced. The following morning those engaged in collecting the opium lay a large poppy leaf on the palm of their left hand, and, having a suitable knife in their right hand, they scrape the then brown concreted juice which has exuded from the incision in each capsule, and transfer it from the knife to the leaf. At every alternate scraping the knife is wetted with saliva by drawing it through the mouth to prevent the half-dried juice adhering to it. When a mass of sufficient size has been collected to form a cake or lump it is enveloped in poppy leaves and put for a short time in the shade to harden. This opium is purchased by the merchants from the cultivators, and packed in bags in which the chaffy fruits of a species of *Rumex* are placed to prevent the lumps from sticking together ; after which the bags are sealed and forwarded chiefly to Smyrna, but some opium is also sent to Constantinople.

General Characters.—In rounded, irregularly-formed, or flattened masses, varying in weight from an ounce to six pounds or more, but commonly from about eight ounces to two pounds; usually covered with portions of poppy leaves, and scattered over with the reddish-brown chaffy fruits of a species of *Rumex*. When fresh it is plastic, and internally somewhat moist, coarsely granular in appearance, and small shreds of the epicarp of the poppy capsule are commonly observable in its substance ; in colour it is reddish- or chestnut-brown. By keeping it becomes harder, and darkens in colour to blackish-brown, or even quite black if kept for many years. The odour is strong, peculiar, and narcotic, and the taste nauseously bitter. This, the official opium, is, as a general rule, the best kind of opium, yielding on an average a larger proportion of the alkaloid morphine, which, being by far the most important constituent of opium, is used as a test of its quality (see page 33). The percentage of morphine varies however in different specimens. Thus, when good and dried, it yields from about 10 to 15 per cent. of morphine, or sometimes more.

Adulterations.—Various substances have been used for the adulteration of the above kind of opium, such as sand, pounded poppy capsules, tragacanth, gum arabic, molasses, starch, sugar, pulp of figs or apricots, &c. Stones, bullets, shot, bits of clay, and other foreign matters may also be not unfrequently found imbedded in the masses. The only reliable test of its purity is, as already stated, the proportion of morphine it yields.

Varieties of Opium.—Including the official opium as above described, the following are the more important varieties which have been usually distinguished by pharmacologists,—*Smyrna, Constantinople, Egyptian, Persian,* and *East Indian.* These must be briefly referred to ; but all other kinds of opium, namely, those prepared in China, and in various parts of Europe, Australia, and the United States of America, require here no special description.

　1. *Asia Minor Opium.*—Under this head we include the two kinds distinguished above as *Smyrna* and *Constantinople*, or commonly as *Turkey Opium.* We include both Smyrna and Constantinople under one head, because, as described in speaking of the *Production, Extraction, and Collection of Opium*, we have seen that both are produced in the same parts of Asia Minor and prepared in a similar manner, the opium thus collected being then forwarded to Smyrna and Constantinople, and thence exported to Europe and America ; hence the names by which they are more commonly distinguished. Asia Minor opium is that which alone is official, except for obtaining the alkaloids, etc., as before stated, and to which the characters above given apply. In its physical characteristics it is more especially distinguished by being enveloped in poppy leaves and scattered *rumex* fruits, and by its colour gradually changing to blackish-brown or even black by keeping. It is not only the official kind, but by far the more important commercial variety in this country and other parts of Europe, and in the United States.

　2. *Egyptian Opium.*—This kind of opium is derived, like that from Asia Minor, from *Papaver somniferum*, var. *glabrum*, Boissier. It is usually found in flattish or plano-convex cakes, from three to four inches in diameter, and covered with pieces of poppy leaves, but *no rumex fruits* are seen on the surface. It varies much in consistence, being hard and dry or soft and plastic. It is distinguished from Asia Minor opium by its dark liver-brown colour, somewhat resembling hepatic aloes, and by not blackening by keeping ; its odour also is less strong, and somewhat musty ; as a rule it is very inferior to Asia Minor opium, but its quality is subject to much variation. Its importation is very irregular, and at present is but rarely seen in commerce.

　3. *Persian Opium.*—This is the produce of *Papaver*

somniferum, var. *album*, Boissier. It has been found in various forms, thus, rarely, in somewhat flattened cylindrical sticks, but far more commonly in short roundish cones, flat circular cakes, or irregular rounded masses. The sticks form the Trebizond opium of Pereira ; each stick is wrapped in smooth shiny paper, and tied with cotton, and about six inches long. The other forms are either wrapped in paper or covered with broken poppy leaves and stalks. As a general rule it has a firm consistence, a light brown colour, a good opiate odour, and its external surface, as well as its substance, present an oily appearance. This oily character is very distinctive of Persian opium amongst the ordinary commercial kinds of opium, and is caused by such opium being collected with a knife or flat scraper rubbed over, as well as the fingers of the collector, with linseed oil. Its supply is uncertain, but in some years it arrives in large quantities. It is generally superior in quality to Egyptian opium, but on an average inferior to the Asia Minor kind, although in some cases it has been found to yield more than 10 per cent. of morphine.

4. *East Indian Opium.*—This is derived from *Papaver somniferum*, var. *album*, Boissier, like that of the Persian kind. It is found in various forms, and generally of very inferior quality, although some yield a good percentage of morphine. But as it is never seen in European commerce, being essentially prepared for use in China, and to some extent in India, it is unnecessary to refer to it further than to say that it is exported to China in enormous quantities, no less on an average annually than about eight millions of pounds, representing a market value of about as many pounds sterling. Hence its importance as a source of revenue in our Indian Empire.

Principal Constituents.—Opium is remarkable for the number of peculiar principles which it contains, or that may be obtained from it as secondary or derivative compounds. Other constituents are *mucilage, glucose, pectin, caoutchouc, wax, odorous principle, colouring matter*, &c., but there is *neither starch nor tannic acid* in opium. Good dried opium yields from 4 to 8 per cent. of ash. *Lactic acid* is also a primary or secondary constituent of opium.

The peculiar principles are : *meconic acid* ; two neutral substances—*meconin* or *opianyl* and *meconosin* ; and numerous alkaloids, as follows :—*morphine, codeine, narcotine, pseudomorphine* or *phormine, thebaine* or *paramorphine, narceine, papaverine, rhœadine, lanthopine, laudanine, codamine, protopine, hydrocotarnine, meconidine, cryptopine, laudanosine, gnoscopine,* and *deuteropine.* Of these alkaloids by far the most important is morphine, and hence, as we have noticed, the quality of opium is estimated by the yield per cent. of this alkaloid. The last eleven alkaloids are only present in minute proportion, and deuteropine is not known in a state of purity.

Besides the above peculiar principles, a number of derivative or secondary compounds have been obtained from opium, of which *apomorphine* is the most important in a medicinal point of view.

Medicinal Properties.—Opium is, as Pereira says, the most important and valuable remedy of the whole Materia Medica. Its primary effect is stimulant; its secondary effects hypnotic, anodyne, and antispasmodic ; and in overdoses it is a powerful narcotic poison.

Morphine generally resembles opium in its effects, but it is much less stimulant. Codeine is hypnotic and stimulant, but more excitant than morphine. Narcotine is tonic and antiperiodic, and in large doses diaphoretic. Thebaine is feebly hypnotic, and excitant like strychnine. Narceine and papaverine are feebly hypnotic. Cryptopine is hypnotic and excitant. Meconin is feebly hypnotic. Apomorphine is a powerful emetic. Meconic acid is almost inert. Of the properties of the other principles nothing definite is known.

Doses.—½ grain to 3 grains of Opium. Of Morphine, ¼ to ½ a grain. Of Codeine, ¼ grain to 2 grains. Of Narcotine, 1 grain to 3 grains as a tonic ; 5 to 20 grains as

an antiperiodic. Of Apomorphine, as an emetic, 0·15 to
0·3 grains given by the mouth; or, used hypodermically, 0·05
grain to 0·1 grain.

Official Preparations of Opium.

Confectio Opii. *Dose.*—5 to 20 grains.
Emplastrum Opii.
Enema Opii.
Extractum Opii. *Dose.*—½ grain to 2 grains.
Extractum Opii Liquidum. *Dose.*—10 to 40 minims.
Linimentum Opii.
Pilula Ipecacuanhæ cum Scilla. *Dose.*—5 to 10 grains.
Pilula Plumbi cum Opio. *Dose.*—3 to 5 grains.
Pilula Saponis Composita. *Dose.*—3 to 5 grains.
Pulvis Cretæ Aromaticus cum Opio. *Dose.*—10 to 40
grains.
Pulvis Ipecacuanhæ Compositus. *Dose.*—5 to 15 grains.
Pulvis Kino Compositus. *Dose.*—5 to 20 grains.
Pulvis Opii Compositus. *Dose.*—2 to 5 grains.
Suppositoria Plumbi Composita. Each suppository con-
ains one grain of opium.
Tinctura Camphoræ Composita. *Dose.*—15 minims to
1 fluid drachm.
Tinctura Opii. *Dose.*—5 to 40 minims.
Tinctura Opii Ammoniata. *Dose.*—½ to 1 fluid drachm.
Trochisci Opii. *Dose.*—1 to 6 lozenges.
Unguentum Gallæ cum Opio.
Vinum Opii. *Dose.*—10 to 40 minims.

3. Official Alkaloids of Opium and their Salts; and Meconic Acid.

Besides crude opium and its official preparations, Acidum
Meconicum and the following alkaloids or their salts are
also official :—
Apomorphinæ Hydrochloras.
Codeine. *Dose.*—¼ to 2 grains.

Morphinæ Acetas.   *Dose.*—$\frac{1}{8}$ to $\frac{1}{2}$ grain.
Morphinæ Hydrochloras.   *Dose.*—$\frac{1}{8}$ to $\frac{1}{2}$ grain.
Morphinæ Sulphas.   *Dose.*—$\frac{1}{8}$ to $\frac{1}{2}$ grain.

*Official Preparations of Meconic Acid, and of the Salts of
the Alkaloids of Opium.*

1. Of ACIDUM MECONICUM :—
Liquor Morphinæ Bimeconatis.   *Dose.*—5 to 40
minims.

2. Of APOMORPHINÆ HYDROCHLORAS :—
Injectio Apomorphinæ Hypodermica.   *Dose, by
subcutaneous injection.*—2 to 8 minims.

3. Of MORPHINÆ ACETAS :—
Injectio Morphinæ Hypodermica.   *Dose, by subcutaneous injection.*—Commencing with from 1
to 2 minims.
Liquor Morphinæ Acetatis.   *Dose.*—10 to 60
minims.

4. Of MORPHINÆ HYDROCHLORAS :—
Liquor Morphinæ Hydrochloratis.   *Dose.*—10 to
60 minims.
Suppositoria Morphinæ.   Each suppository contains half a grain of hydrochlorate of morphine.
Suppositoria Morphinæ cum Sapone.   Each suppository contains half a grain of hydrochlorate
of morphine.
Tinctura Chloroformi et Morphinæ.   *Dose.*—5 to
10 minims.
Trochisci Morphinæ.   Each lozenge contains one
thirty-sixth of a grain of hydrochlorate of morphine.   *Dose.*—1 to 6 lozenges.
Trochisci Morphinæ et Ipecacuanhæ.   Each
lozenge contains one thirty-sixth of a grain of
hydrochlorate of morphine.   *Dose.*—1 to 6
lozenges.

2. PAPAVER RHŒAS, *Linn.*
Red Poppy.

(Bentley and Trimen's ' Medicinal Plants,' vol. i. plate 19.)

Habitat.—A common weed throughout Europe except Scandinavia. It is abundant in England and Ireland, but less so in Scotland. It also extends through Asia Minor to North-West India.

Official Part and Name.—RHŒADOS PETALA :—the fresh petals. From indigenous plants.

Rhœados Petala.
Red Poppy Petals.

General Characters.—The fresh petals have a bright scarlet colour except at the base, where they are often nearly black ; they are thin, and unequal in size. They have a strong narcotic odour, and slightly bitter taste. When dried with the greatest care, they become brownish-violet-red, crumple up, lose their odour, but still retain a slight bitterness.

Principal Constituents.—Their principal constituent is the colouring matter, in the proportion of about 40 per cent. This colouring matter is readily exhausted by water and rectified spirit ; acids diminish its intensity, while alkalies cause it to become nearly black. It is said to be composed of two amorphous acids, which have been termed by Leo Meier, their discoverer, *rhœadic* and *papaveric*. None of the peculiar alkaloids of opium can be traced in the petals.

Medicinal Properties.—They have been supposed to possess, especially when fresh, very slight narcotic properties, but are now only employed for their beautiful colour.

Official Preparation.—Syrupus Rhœados.

ORDER 6.—CRUCIFERÆ.

1. COCHLEARIA ARMORACIA, *Linn.*
Horseradish.

(Bentley and Trimen's ' Medicinal Plants,' vol. i. plate 21.)

Habitat.—Apparently indigenous to Eastern Europe, extending from the Caspian through Russia and Poland to Finland. In Britain and other parts of Europe, except the extreme South, although apparently wild in many localities, it has probably been introduced.

Official Part and Name.—ARMORACIÆ RADIX :—the fresh root. From plants cultivated in Britain, and most active in the autumn and early spring before the leaves have appeared.

Armoraciæ Radix.
Horseradish Root.

General Characters.—Nearly cylindrical, except at the upper end or crown, where it is enlarged and conical, and marked in an annulated manner by the scars of fallen leaves (*fig.* 2, *c*) ; it is also frequently divided near this part into two or more short branches, each being surmounted by a tuft of leaves (*fig.* 2, *d*). In diameter it varies from half to one inch or more, and it is commonly a foot or more in length. Externally it is pale yellowish-white or brownish-white; and whitish and fleshy within. Its taste is very pungent, more especially in spring and autumn, accompanied by bitterness or sweetness, according to the soil in which it is grown, the manner in which it is cultivated, and the season in which it is collected. It is inodorous, except when bruised or scraped, when it is highly pungent, frequently causing sneezing and a flow of tears.

We have already, under the head of 'Aconiti Radix,' given in a tabulated form the distinctive characters of horseradish root and aconite root, in consequence of the fatal cases of poisoning that have occurred from the latter root having been substituted as a condiment for the former. (See page 16.)

Principal Constituents.—The active constituent is a *volatile oil*, which may be obtained in the proportion of about o·2 per cent. by the distillation of fresh horseradish root with water. This oil is identical with the volatile oil obtained from black mustard seeds, and like it does not pre-exist, but is developed from the action of an albuminous body (*myrosin*) and myronate of potassium (*sinigrin*) in the presence of water (see page 46). The presence of myrosin in horseradish root has, however, never been experimentally demonstrated.

Medicinal Properties.—Internally administered, horse-radish root is stimulant, diuretic, and diaphoretic; externally applied it is rubefacient or even vesicant. When chewed it acts as a sialagogue. Its chief use is, however, not as a medicine, but as a condiment, and when partaken of in moderation, it increases the appetite and promotes digestion.

Official Preparation.—Spiritus Armoraciæ Compositus.

This preparation is most active when prepared in the autumn, or in the early spring before the leaves have appeared.

---

2. BRASSICA NIGRA, *Koch.*

Black, Brown, or Red Mustard.

Synonym.—SINAPIS NIGRA, *Linn.*

(Bentley and Trimen's ' Medicinal Plants,' vol. 1, plate 22.)

Habitat.—It is found throughout England and the South of Scotland, but must be regarded as a doubtful native. It grows throughout Europe, except the extreme North-East parts. It also occurs in North Africa, in Asia Minor, and other parts of Asia, and is naturalised in both North and South America.

Official Parts or Products and Names.— 1. SINAPIS NIGRÆ SEMINA :—the dried ripe seeds. From plants cultivated in Britain. 2. OLEUM SINAPIS :—the oil distilled with water from black mustard seeds after the expression of the fixed oil. 3. SINAPIS :—black mustard seeds and white mustard seeds powdered and mixed.

1. Sinapis Nigræ Semina.

Black Mustard Seeds.

The official name of *Sinapis* is the former generic name, which is derived from the Greek word σίναπι, *mustard*.

General Characters. — Black mustard seeds are also termed *Brown* and *Red Mustard Seeds* ; they are scarcely half the size of white mustard seeds, or not more than one twenty-fifth of an inch in diameter, and are roundish in form. In colour they are dark reddish-brown or greyish-brown externally, the testa being netted with minute pits ; internally they are yellow. They are hard, and their powder has a yellowish-green colour. When entire they are inodorous, and even when powdered dry ; but when triturated with water they exhale a strong pungent odour, so as to affect the eyes, or even to cause a flow of tears. Their taste is somewhat bitter at first, but this is immediately followed by pungency.

Principal Constituents.—Black mustard seeds contain about 25 per cent. of a *fixed oil* ; an albuminous body resembling the emulsin of almonds, termed *myrosin*; and a crystallisable substance called *myronate of potassium* or *sinigrin,* and other unimportant constituents, but the ripe seeds contain *no starch.* The volatile oil, upon which the activity of the seeds depends, does not exist in a free state, but only as a constituent of sinigrin, which substance is a compound of *sulphocyanate of allyl* or *mustard oil, bisul-*

phate of potassium, and *sugar* (dextroglucose); and which is separated into the three substances of which it is composed when dissolved in water in contact with myrosin (*see* Oleum Sinapis).

Medicinal Properties.—Black mustard seed is a powerful stimulant, and also emetic and diuretic; it is chiefly employed, both externally and internally, in the form of the official mustard (*Sinapis*); and externally as a rubefacient in the form of the official *Charta Sinapis.*

Official Preparations.

Oleum Sinapis (*which see*). | Sinapis (see *Sinapis*).

2. Oleum Sinapis.

Oil of Mustard.

Production.—The volatile oil, as we have seen (*Principal Constituents of Black Mustard Seeds*) does not exist in a free state, but as a constituent of sinigrin, and is liberated whenever sinigrin is dissolved in water, and brought into contact with myrosin, as is the case when prepared according to the directions of the British Pharmacopœia. The temperature of the water used in its distillation should, however, not exceed 122° Fahr., as a much higher degree of heat will prevent the formation of the oil.

General Characters.—Oil of mustard is colourless or pale yellow, boils at about 298° F. (147·8° C.), and has a specific gravity of from 1·015 to 1·020 at 60° F. It is readily soluble in alcohol or ether, but only very slightly so in water. It has an exceedingly pungent odour, and a very acrid burning taste. The pungent taste and odour, and the medicinal properties generally of black mustard seeds are due to this oil. As already noticed (page 46), chemists regard it as the *sulphocyanate of allyl.*

Medicinal Properties.—When applied to the skin in a pure state it produces almost instant vesication, but when dissolved in alcohol, or in the form of the official Compound Liniment of Mustard, it acts as a powerful rubefacient.

Official Preparation.—Linimentum Sinapis Compositum.

3. Sinapis.

Mustard.

The official mustard being a mixture of black mustard seeds and white mustard seeds, its description is postponed until after the latter seeds have been noticed (see page 49).

---

3. BRASSICA ALBA, *Hook. fil. & Thomp.*

White Mustard.

Synonym.—Sinapis alba, *Linn.*

(Bentley and Trimen's ' Medicinal Plants,' vol. i. plate 23.)

Habitat.—This plant occurs throughout Europe, more especially in the south, where it is probably a native. It is a frequent weed in England, but has even less claim than *Brassica nigra* to be regarded a British plant. It is also found in Algeria, Asia Minor, and China.

Official Parts and Names.—1. Sinapis Albæ Semina :— the dried ripe seeds. From plants cultivated in Britain. 2. Sinapis :—black mustard seeds and white mustard seeds powdered and mixed.

1. Sinapis Albæ Semina.

White Mustard Seeds.

General Characters.—About one-twelfth of an inch in diameter, roundish, of a pale yellow colour, and with a very finely pitted reticulated testa. They have a hard texture, and internally are yellow and oily. Inodorous when entire or powdered, and almost so even when triturated with water. Taste pungent, but less so than black mustard seeds.

Principal Constituents.—When submitted to pressure they yield a similar *fixed oil* to that derived from black mustard seeds under the same circumstances. They also contain a crystalline principle termed *sulpho-sinapisin* or *sinalbin*, which is a compound of *sulphocyanate of acrinyl, sulphate of sinapine*, and *glucose*; and *myrosin*; but *no starch.* When sinalbin is dissolved in water at ordinary temperatures in contact with myrosin, it is resolved into its three constituent substances, and it is to the sulphocyanate of acrinyl that white mustard seeds owe their pungency and medicinal properties. This principle, like the active principle of black mustard seeds, does not therefore exist in a free state in white mustard seeds, but only as a constituent of sinalbin; but it cannot be obtained, like the volatile oil of mustard, by distillation, as sulphocyanate of acrinyl is a fixed acrid oily substance.

Medicinal Properties.—Stimulant, but milder than those of black mustard seeds. Also emetic, diuretic, and laxative. They are chiefly employed in the form of the official mustard (*Sinapis*), which see.

Official Preparation.—Sinapis.

2. Sinapis.

Mustard.

Production.—As we have already noticed, the official *mustard*, or *flour of mustard* as it is often termed, is a mixture of powdered black and white mustard seeds. Commercial mustard is, in like manner, a mixture of these seeds in varying proportions. A mixture of the two seeds forms the best mustard, because in black mustard seeds there is scarcely sufficient myrosin to decompose the whole of the sinigrin, while in white mustard seeds the amount is more than is required to decompose the sinalbin.

E

General Characters.—A greenish-yellow powder, having an acrid, bitterish, oily, pungent taste. Scentless when dry, but exhaling when moist a pungent, penetrating, peculiar odour, which is very irritating to the nostrils and eyes.

Adulterations.—Wheaten flour, or starch coloured by turmeric, is frequently added to mustard as an adulteration, the loss of pungency being compensated for by the addition of some powdered capsicum fruit. This adulteration is readily detected by the following test given in the pharmacopœia : a decoction cooled is not made blue by tincture of iodine.

Medicinal Properties.—Emetic, diuretic, and stimulant when given internally. Externally applied, it is a valuable rubefacient ; and if kept long in contact with the skin, it is vesicant. As a condiment in moderate use, it promotes the appetite, and assists the assimilation of substances difficult of digestion.

Official Preparations.

Cataplasma Sinapis. | Charta Sinapis.

ORDER 7.—CANELLACEÆ.

CANELLA ALBA, *Murray.*
Wild Cinnamon.

(Bentley and Trimen's ' Medicinal Plants,' vol. i. plate 26.)

Habitat.—Native of Jamaica, Cuba, and other of the West Indian Islands, the Bahamas, and the South of Florida.

Official Part and Name.—CANELLÆ CORTEX :—the bark deprived of its corky layer and dried.

Canellæ Cortex.
Canella Bark.

Collection and Commerce.—Incisions are first made through the bark to the wood at intervals of commonly about a foot.   The corky layer is then removed by beating with a stick, and by a further beating the bark is separated from the wood, peeled off, and dried in the shade.   It is collected in the Bahama Islands, and shipped to Europe from Nassau in New Providence.

General Characters.—In more or less twisted quills or irregular pieces, which are generally broken longitudinally, and vary in length from two inches to one foot or more, in width from half an inch to about two inches, and in thickness from one-twelfth to one-eighth of an inch.   Externally, it has a pale orange-brown or buff colour ; is marked by slight transverse wrinkles and evident rounded depressions or scars ; and sometimes the remains of the corky layer may be seen here and there as silvery-grey patches.   Internally, it is paler coloured, being whitish or yellowish-white, and very finely striated longitudinally or quite smooth.   Its fracture is short and granular, the fractured surface showing distinctly the two layers of which canella bark is essentially composed. It has an agreeable odour, somewhat resembling a mixture of cinnamon and cloves ; and a bitter pungent acrid taste.

Principal Constituents.—Its more important constituent is a *volatile oil,* in the proportion of from 0·75 to 0·90 per cent.   There is also about 8 per cent. of *mannite,* and a *bitter principle* which has not been isolated.   *It does not contain any tannic acid,* hence its aqueous infusion is not blackened by a persalt of iron.   By the latter test, canella bark may be readily distinguished both from true *Winter's Bark,* obtained from *Drimys Winteri,* Forster, a tree of the order Magnoliaceæ ; and from the bark of *Cinnamodendron corticosum,* Miers, of the order Canellaceæ, and commonly sold in this country as *Winter's Bark,* because an infusion of either of these barks is blackened by a persalt of iron.

Medicinal Properties.—Aromatic stimulant and tonic. Also used in the West Indies as a condiment.

Official Preparation.—Vinum Rhei.

ORDER 8.—POLYGALACEÆ.

1. POLYGALA SENEGA, *Linn.*

Rattlesnake Root. Senega Snake Root.

(Bentley and Trimen's 'Medicinal Plants,' vol. i. plate 29.)

Habitat.—A native of North America, extending from the Northern parts of Canada and through the United States as far southward as North Carolina and Tennessee.

Official Part and Name.—SENEGÆ RADIX :—the dried root.

Senegæ Radix.

Senega Root.

Collection.—It is collected in the Southern and Western parts of the United States.

General Characters.—Varying in length, but usually about four inches, and from one-fifth to one-third of an inch or more thick ; and enlarged at the upper end into an irregular thick knotty tuberosity or crown, which bears the remains of numerous wiry aerial stems and scaly rudimentary leaves. Below the crown, the root is more or less twisted, curved, or spiral, and ultimately divides into two or more branches ; and has in most cases a keeled-shaped ridge running along its whole length. The bark is thick, yellowish-grey or brownish-yellow externally, whitish within ; more or less wrinkled, somewhat knotted ; transversely cracked, so as to be partially annulated, horny, translucent ; and enclosing an irregularly-formed, whitish, central woody column, about as thick as itself. Fracture short and brittle. Odour of bark slight, but peculiar, rancid, and unpleasant ; and its taste is at first sweetish, but subsequently sourish, very acrid, and causing a flow of saliva. The wood is tasteless and inodorous.

Principal Constituents.—Senega root contains about 5

per cent. of *polygalic acid* or *senegin*, to which its properties are essentially due; also a *fixed oil*, a little *volatile oil, sugar*, &c. Senegin is closely allied to, if not identical with, saponin.

Substitutions and Adulterations.—The roots of other species or varieties of *Polygala* are sometimes substituted for the official root. Such are the roots of *Polygala Boykinii*, Nuttall, and those of a northern variety of the true *Polygala Senega*. The former has no keel, and its central woody column is cylindrical; the latter is frequently keeled, like the true root, but it has a regular cylindrical column. Both roots are said to contain *polygalic acid*, but not in so large a proportion as the true root.

Senega root is not intentionally adulterated ; but from ignorance, or carelessness in its collection, other roots and rhizomes may be frequently found mixed with it. Thus, American Ginseng root, from *Panax quinquefolium*, Linn., is that most commonly seen; it is readily distinguished by its more or less fusiform shape, greater thickness, by the absence of any keel, and other characters. The roots of *Gillenia trifoliata*, Moench, and other species of *Gillenia*; and the rhizomes of *Cypripedium pubescens*, Willd.; and of *Cynanchum Vincetoxicum*, R. Brown, have also been noticed. All of these are readily distinguished by the absence of a keel, their different tastes and odours, and by other marked characters.

Medicinal Properties.—Chiefly employed as a stimulant expectorant, but also as a diaphoretic and diuretic in moderate doses; it is emetic and cathartic in large doses. It is useless as a remedy for the bites of venomous reptiles.

Official Preparations.

Infusum Senegæ. | Tinctura Senegæ.

2. KRAMERIA TRIANDRA, *Ruiz & Pavon.*
Peruvian Rhatany.

(Bentley and Trimen's ' Medicinal Plants,' vol. i. plate 30.)

Habitat.—Peru and Bolivia.

3. KRAMERIA IXINA, *Linn.*, var. GRANATENSIS, *Triana.*
Savanilla or New Granada Rhatany.

Synonym.—KRAMERIA TOMENTOSA, *St. Hil.*

Habitat.—New Granada, British Guiana, and some parts of Brazil.

Official Part and Name.—KRAMERIÆ RADIX:—the dried root of the above two species of Krameria.

Krameriæ Radix.
Rhatany Root.

A. PERUVIAN, PAYTA, or RED RHATANY.

Collection.—As its common name implies, it is collected in Peru, more especially in the districts bordering on Lima ; but also, to some extent, in the northern part of Peru, and is then shipped from Payta, hence it is sometimes known as *Payta Rhatany.*

General Characters.—Peruvian Rhatany is either in long sub-cylindrical branched or unbranched pieces, varying in diameter from a quarter to half an inch or more ; or more commonly it consists of a short portion of varying thickness, but sometimes as large as a man's fist, and usually much knotted, from which a variable number of short, stumpy, more or less broken branches, of different sizes, arise. The *bark*, which is readily separable from the wood beneath varies in thickness from about one-twentieth to one-tenth of an inch (*fig.* 10, *a*), it is rough and scaly except in the smaller pieces, dark reddish-brown on its outer surface (hence the name of *Red Rhatany*, applied to this root), and bright brownish-red on its inner ; breaks with a somewhat fibrous fracture, and is tough and difficult to powder. The *woody axis* is hard, and of a brownish- or reddish-yellow colour. The *bark* has a strongly astringent taste, and when chewed tinges the saliva red, but no odour ; the *wood* is nearly tasteless and inodorous.

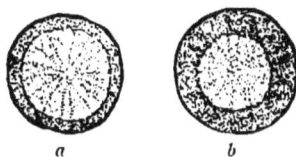

Fig. 10.—*a.* Transverse section of *Peruvian Rhatany Root. b.* Transverse section of *Savanilla Rhatany Root.* (After Maisch.)

Principal Constituents.—About 20 per cent. of a form of tannic acid, to which its properties are essentially due, closely allied to catechu-tannic acid, and called *ratanhia-tannic acid* or *krameria-tannic acid* ; and *ratanhia-red.* There is no gallic acid in rhatany root.

B. SAVANILLA or NEW GRANADA RHATANY.

Collection.—It is collected in New Granada, in the valley between Pamplona and the Magdalena.

General Characters.—Savanilla Rhatany is less knotty and irregular than that of the Peruvian kind ; the pieces are also shorter and not so thick, being commonly from four to eight inches in length, and from one-fifth to nearly half an inch in thickness. It is also well characterised by its thick bark (*fig.* 10, *b*), which is from one-fourth to one-third of that of the woody axis, while that of Peruvian rhatany (*fig.* 10, *a*) rarely exceeds one-sixth of the diameter of the wood, and is commonly less. This bark is firmly adherent to the woody axis, its outer surface is smooth, of a dull pale purplish-brown colour, and is marked by slight longitudinal furrows, and usually, at irregular intervals, by deep narrow transverse cracks. It is also less fibrous and tough than that of Peruvian rhatany, and may therefore be more readily powdered. Like Peruvian rhatany the bark has an intensely astringent taste, but no odour ; and the wood is nearly taste-less and inodorous.

Principal Constituents.—It owes its properties, like that of Peruvian rhatany, to some form of *tannic acid*, but the action of reagents is somewhat different on the two roots, and hence their tannic matter is not of the same character. Thus, if a thin section of Peruvian rhatany be moistened by a solution of a ferrous salt, it becomes greyish ; while a sec-tion of Savanilla rhatany under the same circumstances assumes an intense violet colour.

Medicinal Properties.—Similar to those of catechu, being a powerful astringent and also tonic like other drugs of this class. The two kinds of rhatany are identical in their action, the Savanilla rhatany being probably the better of the two.

Official Preparations.

Extractum Krameriæ.	Pulvis Catechu Compositus.
Infusum Krameriæ.	Tinctura Krameriæ.

The two above kinds of rhatany may be used indif-ferently in the official preparations.

C. Other Kinds of Rhatany.

(*Not Official.*)

Besides the two official kinds of rhatany root, others have been also described by pharmacologists, but only one of these is commonly found in commerce ; it is not un-frequently seen in the London market. It is said to be derived from *Krameria argentea*, Martius, and is collected in the North-Eastern provinces of the Brazils. It is com-monly known as *Para Rhatany*, but also as *Ceara Rhatany*, and likewise, from its colour, as *Brown Rhatany*. It has a close resemblance to Savanilla rhatany, but is distinguished from it by its darker colour, being brownish-grey, instead of dull pale purplish-brown ; hence its name of *Brown Rhatany* ; and by being in longer and very flexible pieces.

It has a thick bark, which is marked by numerous transverse cracks.

Other species of *Krameria* also yield roots similar in their properties to rhatany, but they are not usually found in commerce ; and Holmes has recently described a very astringent root which appeared in the London market as rhatany, and was imported from. Guayaquil. He believes it to be obtained from a genus nearly allied to *Krameria*.

ORDER 9.—GUTTIFERÆ OR CLUSIACEÆ.

GARCINIA HANBURII, *Hook. fil.*
Siam Gamboge.

Synonym.—GARCINIA MORELLA, *Desrous*, var. PEDICELLATA, *Hanbury*.

(Bentley and Trimen's ' Medicinal Plants,' vol. i. plate 33.)

Habitat.—A native of Cambodia and Cochin China on the East coast of the Gulf of Siam.

Official Product and Name.—CAMBOGIA :— a gum-resin obtained from the above plant.

Cambogia.
Gamboge.

Collection and Commerce.—It is commonly obtained by making a spiral incision into the bark round half the circumference of the trunk, and collecting the juice which then slowly exudes, in a length of bamboo, which is placed at the lower end of the incision for that purpose ; and when hardened is obtained in the form of a roll or cylinder by removing the shell of bamboo. It is produced in Cambodia, and imported by way of Singapore, Bankok, or Saigon.

General Characters and Tests.—In solid or hollow cylindrical pieces or rolls, hence the name of *roll* or *pipe gamboge*. These rolls are longitudinally striated on the surface, vary

from about four to eight inches in length, and from about
one to two inches or somewhat more in diameter, and are
either separate from one another, or more or less agglutinated
or folded together so as to form masses of varying sizes and
forms; they are generally covered by a dirty greenish-yellow
powder.   Gamboge is brittle, breaks with a conchoidal
fracture, the fractured surface being opaque, smooth, glistening, and of a uniform reddish-yellow colour; its powder is
bright yellow.   It has no odour, and but little taste at first,
although ultimately very acrid.   It forms a yellow emulsion
when rubbed with water; is completely dissolved by the successive action of rectified spirit and water; and an emulsion
made with boiling water, and cooled, does not become green
on the addition of solution of iodine.

Besides the above variety of gamboge, it occurs sometimes in large, irregular, more or less flattened lumps or
cakes.   This kind when pure has essentially the same
characters, except as to form and striated surface, as roll
gamboge, but it is more liable to adulteration.

Adulterations.—The ordinary adulterants of gamboge
are rice flour, sand, the powdered bark of the tree, and
pieces of wood and bark ; all these are readily detected by
the characters and tests given above.

Principal Constituents.—It is a mixture of *gum* and
resin, the former in varying proportions of from about 16
to 20 per cent.   The gum, although soluble in water, like
gum acacia, is not identical with it, as the solution is not
precipitated by neutral acetate of lead, nor by perchloride
of iron.   The medicinal properties and colour are due to
the resin, which has a very feeble acid reaction, and hence
is sometimes known as *cambogic acid.*

Medicinal Properties.—It is a valuable drastic and hydragogue cathartic, and to some extent is also anthelmintic.
In excessive doses it is an irritant poison.

Official Preparation.—Pilula Cambogiæ Composita.

ORDER 10.—TERNSTRŒMIACEÆ OR CAMELLIACEÆ.

CAMELLIA THEA, *Link.*
The Tea Plant.

Synonym.—THEA CHINENSIS, *Linn.*

(Bentley and Trimen's 'Medicinal Plants,' vol. i. plate 34.)

Habitat.—Probably a native of Upper Assam, and introduced into China at a very early period from India. It is now cultivated on an enormous scale in China, several parts of India (especially Assam), Japan, and Java.

Official Product and Name.—CAFFEINA :—an alkaloid usually obtained from the dried leaves of Camellia Thea, *Link*, or the dried seeds of Coffea arabica, *Linn.* (see Coffea arabica) by evaporating aqueous infusions from which astringent and colouring matters have been removed.

Caffeina.
Caffeine.

Synonyms.—Caffeia ; Theina ; Guaranina.

The description of caffeine does not come within our province, but we must refer to tea leaves as one of its sources. These leaves form our common Tea, which is commonly distinguished as China Tea.

General Characters of Tea Leaves.—When fresh the leaves are petiolate, two to four inches long, oval or lanceolate acute at the apex or somewhat bluntish and emarginate, irregularly and rather distantly dentate-serrate except at the base ; smooth, shining, and dark green above, smooth or somewhat downy, paler coloured, and with strongly marked

veins below. Odour peculiar but agreeable ; taste somewhat astringent and feebly bitter.

Preparation and Varieties of Tea.—There are numerous
varieties of tea known in commerce, all of which are arranged
under the two heads of Black Teas and Green Teas, both
kinds of which may be prepared from the same plant, the
differences between them depending essentially upon their
mode of preparation. Thus, Green teas are prepared
by drying the leaves as quickly as possible after they are
gathered, and then slightly heating them in a pan over a slow
fire, after which they are rolled separately or in small heaps
with the object of twisting them, and finally dried as quickly
as possible over a fire. Black teas are made from the leaves,
which, after being gathered, are exposed to the air for some
time, and then, after having been tossed about, are placed
in heaps, where they undergo a kind of fermentation, after
which they are heated in a pan over a slow fire for a short
time, then rolled in masses to give them a twisted character ;
then they are again placed in the air, and subsequently submitted a second time, in a shallow pan, to the heat of a
charcoal fire, rolled again, and exposed to the air, and
finally dried slowly over a fire. The differences, therefore,
in the preparation of black and green teas are most marked,
and fully sufficient to account for their difference of colour
and other peculiarities. Thus green teas consist of the
leaves quickly dried after gathering, so that their colour and
other properties are in a great measure preserved ; while
black teas are composed of the leaves dried some time
after having been gathered, and after they have undergone
a kind of fermentation by which their original green colour
is changed to black, and other important changes produced.

Adulterations.—A great part of the green tea which is
exported from China, and consumed in this country and
elsewhere, is coloured artificially with a mixture of finely
powdered prussian blue and gypsum, turmeric being also
sometimes added. Both black and green teas are also

frequently adulterated with the dried and prepared leaves of other plants, such as those of the willow, sloe, elm, &c.; and other adulterations are also practised, but the consideration of these does not come within our province. Tea leaves may, however, be readily distinguished from other leaves by the characters given above.

Principal Constituents.—The properties of tea depend especially upon the presence of a *volatile oil, tannic acid,* and the alkaloid *theine;* the latter being official under the name of *caffeine.* The volatile oil is in the proportion of from ·6 to 1 per cent.; that of tannic acid from 13 to 18 per cent.; and of theine from $1\frac{1}{2}$ to 4 per cent.

Medicinal Properties.—Tea is a nervine stimulant and astringent ; it is not, however, employed as a medicinal agent in this country except in the form of its alkaloid theine or caffeine, and of its salts, as citrate of caffeine, &c., both of which are nervine stimulants. The principal use of tea is to form an agreeable, slightly stimulating, soothing, and refreshing beverage, for which purpose its consumption is enormous, being in Great Britain alone over 140,000,000 pounds yearly.

Official Preparation of Caffeina.

Caffeinæ Citras.   *Dose.*—2 to 10 grains.

The dose of Caffeina is from 1 to 5 grains.

ORDER 11.—MALVACEÆ.

1. ALTHÆA OFFICINALIS, *Linn.*
Marsh Mallow.

(Bentley and Trimen's ' Medicinal Plants,' vol. i. plate 35.)

Habitat.—It is distributed throughout Europe, with the exception of Scandinavia and North Russia. It is also found in Asia Minor, Western Asia, and Algeria. In

England it may be seen in a wild state, more or less, in most of the southern counties.

Part Used and Name.—ALTHÆÆ RADIX :—the root deprived of its brown corky layer and small rootlets, and dried.

(*Not Official.*)

Althææ Radix.
Marsh Mallow Root.

Cultivation and Preparation.—Marsh Mallow is largely cultivated on the Continent, more especially in Bavaria and Würtemburg, and the roots are prepared for use as follows :—In the autumn when the plants are in their second year's growth, the roots are dug up, washed, and scraped so as to remove the outer portions of the bark and the small rootlets, and then dried. This is the condition in which they are usually found in commerce, and preferred for medicinal use; but in some cases the roots are simply washed and dried.

General Characters.—Marsh Mallow root, as prepared by *scraping*, is in sub-cylindrical or somewhat conical pieces, from five to eight inches long, and varying in thickness from that of a common quill to the ring finger. It is whitish externally, deeply and irregularly furrowed longitudinally, marked with the circular scars of the detached rootlets, and more or less covered by projecting liber-fibres. When broken it is seen to consist of a central whitish woody portion, separated by a dark wavy ring from a thick bark ; the central column breaks short, but the bark presents a fibrous appearance. Odour faint and peculiar ; taste mucilaginous and somewhat sweetish. The *unscraped root* has similar characters, but its surface is yellowish-brown, and without any projecting liber fibres ; but sometimes with attached rootlets.

Principal Constituents.—*Mucilage, sugar, starch,* and a neutral crystalline nitrogenous substance termed *asparagin,* are the principal constituents of marsh mallow root. Its properties are essentially due to the mucilage.

Medicinal Properties.—Demulcent and emollient. It is much used on the Continent, but comparatively little in this country.

Preparations.

Syrupus Althææ is the official preparation in the Pharmacopœia of the United States of America ; and in France a favourite preparation is the Pâte de Guimauve.

---

2. GOSSYPIUM BARBADENSE, *Linn.*

Cotton. Sea Island Cotton.

(Bentley and Trimen's ' Medicinal Plants,' vol. i. plate 37.)

Habitat.—Probably a native of the West Indies, but now cultivated over a large part of the warmer regions of the globe, and from it the best long-staple cottons of commerce are obtained.

Official Part and Name.—GOSSYPIUM :—the hairs of the seeds of Gossypium barbadense, *Linn.*; and of other species of Gossypium, from which fatty matter and all foreign impurities have been removed.

Gossypium.

Cotton Wool.

Synonym.—Cotton.

General Characters and Tests.—In white soft filaments, each being from about four-fifths of an inch to more than one and a half inch long according to the staple, and consisting of an elongated tubular cell.   When examined under the microscope in a dried state, it appears as a flattened hollow twisted band, with slightly thickened rounded edges. Inodorous and tasteless. It should be readily wetted by water, to which it should not communicate either an alkaline or acid reaction.   On ignition in air it burns, leaving less than 1 per cent. of ash.

Principal Constituents.—Cotton consists of nearly pure *cellulose*, and from 9 to 10 per cent. of *fixed oil.* When the latter is removed, as in the official cotton, it is known as *absorbent cotton.*

Medicinal Properties.—Cotton as well as the official preparations of Pyroxylin or Gun Cotton, known as Collodium and Collodium flexile, are useful for many purposes in surgery, medicine, and pharmacy.

Official Preparation.—Pyroxylin.—From which Collodium and Collodium Vesicans are prepared; the former being also used for the preparation of Collodium Flexile.

ORDER 12.—STERCULIACEÆ.

THEOBROMA CACAO, *Linn.*

Cocoa or Chocolate Tree.

(Bentley and Trimen's ' Medicinal Plants,' vol. i. plate 38.)

Habitat.—This tree is a native of parts of the Brazils and other northern portions of South America, extending also into Central America as far north as Mexico. It is extensively cultivated throughout the tropics of both the new and old worlds, especially in some of the West Indian Islands.

Official Product and Name.—OLEUM THEOBROMATIS :—a concrete oil obtained by expression and heat from the ground seeds.

Oleum Theobromatis.

Oil of Theobroma.

Synonym.—Cacao Butter.

Before describing the characters of this oil we must refer generally to the seeds from which it is obtained.

Cocoa or Cacao Seeds.—From these seeds both Cocoa and Chocolate are prepared, hence they are sometimes known as *chocolate nuts*. *Cocoa* is either prepared by grinding up the roasted entire seeds between hot cylinders into a paste, which is then mixed with starch, sugar, &c., and formed into the various kinds of cocoa, known as *common cocoa, rock cocoa, soluble cocoa*, &c.; or the roasted seeds, divested of their husks, are broken into small fragments, and constitute *cocoa nibs*. *Chocolate* is prepared from the roasted seeds divested of their husks, and the kernels thus obtained are crushed between heated rollers into a paste. This paste is then mixed with sugar, and some vanilla or cinnamon added for flavouring, or in the inferior kinds some sassafras nuts, cloves, or other aromatic; and usually a small quantity of arnatto, as a colouring agent, is subsequently added, after which it is moulded into cakes, &c.

Both Cocoa and Chocolate, which are so extensively used for the preparation of agreeable and nutritious beverages, &c., are not so stimulating as Tea or Coffee, but they disagree with many persons on account of their fatty or oily nature. To such persons cocoa nibs should be recommended.

The properties of Cocoa seeds are especially due to the alkaloid *theobromine*, which closely resembles the official alkaloid caffeine or theine ; and to the official oil *Oleum Theobromæ*, now to be described.

General Characters of Oleum Theobromæ.—This oil, which is also known as *Cocoa* or *Cacao Butter*, is commonly found in the form of flattened cakes or oblong tablets, averaging half a pound in weight. It has a yellowish colour ; is of about the consistency of tallow, opaque, has a bland agreeable taste, and a pleasant chocolate-like odour. It is unctuous to the touch, but breaks readily and presents a dull waxy fracture. Its specific gravity is about ·960 ; and it usually melts at temperatures between 86° and 95° F. (30° and 35° C.). It does not readily become rancid

F

from exposure to the air. Examined by polarised light, under the microscope, it is seen to consist of minute crystals.

Principal Constituents.—*Stearin, palmitin, olein.*

Medicinal Properties.—Emollient and demulcent. It is chiefly used as a basis for suppositories and pessaries.

Official Preparations.

Suppositoria Acidi Tannici. | Suppositoria Iodoformi.
Suppositoria Hydrargyri. | Suppositoria Morphinæ.
Suppositoria Plumbi Composita.

Series II. *DISCIFLORÆ.*

Order i.—LINACEÆ.

1. LINUM USITATISSIMUM, *Linn.*

Cultivated Flax.

(Bentley and Trimen's ' Medicinal Plants,' vol. i. plate 39.)

Habitat.—The native country of the cultivated Flax cannot now be determined, but it is found in a quasi-wild condition in all the countries where it is grown, but is nowhere known as truly wild.

Official Parts or Products and Names.—1. LINI SEMINA : —the dried ripe seeds. 2. LINI FARINA :—linseed reduced to powder. 3. OLEUM LINI :—the oil expressed in Britain without heat from linseed.

1. Lini Semina.

Linseed.

Cultivation and Commerce.—Some Linseed is produced in Britain, but the Flax plant is less cultivated here than

formerly. Our chief supplies are derived from Russia
and India, more especially from Russia; but some seed is
also obtained from other countries, as Germany, Holland,
&c. The English seed, as a general rule, is regarded as the
best.

General Characters.—Linseed or Flax seed is small, but
varying in length from about one-sixth to one-fourth of an
inch, that from warm countries being the larger. The
seeds are ovoid, more or less flattened, somewhat obliquely
pointed at one end; brown, smooth, and shining on their
outer surface, and yellowish-white internally. They are
inodorous, but have a mucilaginous, oily, and slightly bitter
taste.

Principal Constituents.—In an unripe state linseed contains starch, but when ripe this has been converted into
mucilage. A decoction of linseed when cold does not therefore become blue by the addition of solution of iodine. The
official *fixed oil* (see Oleum Lini) is the most important constituent ; it forms nearly one-third of its weight.

Adulterations.—Linseed is frequently found contaminated
by other seeds and cereal grains. These, when of a starchy
nature, may readily be distinguished by the cooled decoction
in such cases becoming blue by solution of iodine ; and
also, as a general rule, these and other seeds may be known
by other marked characters.

2. Lini Farina.

Linseed Meal.

In the British Pharmacopœia of 1867 the cake left after
the expression of the fixed oil, commonly known as *oil-cake*,
when powdered, formed the official linseed meal ; but in
the present pharmacopœia the meal containing the oil—
that is, simply the powdered seeds—*is official.* This latter
is to be preferred to the former ; but as it soon becomes
rancid when exposed to the air by the formation of fatty

F 2

acids from the oxidation of its contained oil, it should never be long kept. It has a greyish-brown colour, and oily character.

3. Oleum Lini.

Linseed Oil.

When obtained by expression without heat from the seeds, as officially directed, less than 20 per cent. of oil is obtained. Hence, in order to increase the yield of oil, the seeds are commonly pressed after being heated, the amount then obtainable being from about 25 to 30 per cent. according to the quality of the seed.

General Characters and Composition.—The best oil is derived by expression without heat. This is viscid, has a yellow colour, faint odour, and bland oleaginous taste. It gradually thickens by exposure to the air. Its specific gravity is about 0·932, and it congeals at −4° Fahr. Expressed with heat the oil has a dark yellowish-brown colour, disagreeable odour, and acrid taste. The especial characteristic of linseed oil is its drying into a transparent varnish when exposed to the air. Its chief constituent is *linolein*, which by exposure to air is chiefly converted into *linoxyn*.

Medicinal Properties.— Linseed is demulcent and emollient, and in the form of an infusion is extensively employed. Linseed meal is a valuable emollient application in the form of a poultice. Linseed oil is also a very useful external emollient application to burns and scalds.

Official Preparations.

1. Of LINI SEMINA :—

    Infusum Lini. | Lini Farina.
            Oleum Lini.

2. Of LINI FARINA :—

    Cataplasma Carbonis. | Cataplasma Lini.
    Cataplasma Conii.   Cataplasma Sinapis.
            Cataplasma Sodæ Chlorinatæ.

3. Of OLEUM LINI :—   ?

There are no official preparations, but the popular application to burns and scalds known as *Carron Oil*
is a mixture of equal parts of Linseed Oil and
Solution of Lime (Lime Water).

---

2. ERYTHROXYLON COCA, *Lamarck*.

Coca Plant. Ipadu.

(Bentley and Trimen's 'Medicinal Plants,' vol. i. plate 40.)

Habitat.—It is scarcely known in a wild state at the
present time, though doubtless originally a native of some
of the same districts in which it is now cultivated.

Official Part and Name.—COCA :—the dried leaves.

Coca.

Coca.

Synonym.—Cuca.

Cultivation, Preparation, and Commerce.—The Coca
plant is very largely cultivated in Peru, Columbia, and
Bolivia, and to some extent also in parts of Brazil, the
Argentine States, and other parts of South America. The
plantations of coca are called *cocals*, the largest and most
productive being in the province of La Paz in Bolivia. The
leaves appear to be always obtained from cultivated plants,
and much care should be taken in their gathering, drying,
and preservation, as the activity of coca depends in a great
degree on its mode of preparation. It is chiefly exported
from Lima. Coca of commerce varies very much in quality,
and care should be taken to obtain it, as far as possible,
with the characters described below.

General Characters.—In commercial specimens of coca
the leaves are either separate from one another, or more

commonly, they are loosely pressed together into masses.
Coca should present the following characters :—Leaves
shortly-stalked (*figs.* 11 and 12), one to two inches or more
in length, flat, oval or lanceolate, of varying thickness, entire,
usually obtuse and emarginate, and with a little point in
the notch (*fig.* 11), quite smooth ; midrib prominent, with
numerous faint freely-anastomosing lateral veins, and on
each side of the midrib a curved line extends from base to

FIG. 11.—A portion of a branch of *Erythroxylon*
Coca, bearing leaves and flowers.

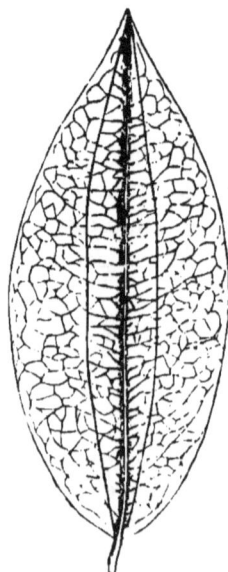

FIG. 12. — Leaf of *Ery-*
throxylon Coca.

apex (*fig.* 12) ; green above, somewhat paler beneath,
more especially between the curved line and the midrib.
Odour faintly aromatic and somewhat tea-like, particularly
when bruised ; taste somewhat aromatic and bitter. Coca
leaves of commerce are commonly more or less broken, and
frequently yellowish-green, yellowish-brown, or brown.

Principal Constituents.—Cocaine a colourless inodorous
bitter crystalline alkaloid, soluble in alcohol or ether, and
also in water; *hygrine*, a volatile oily odoriferous liquid,

with a strong alkaline reaction ; and *coca-tannic acid*. By far the more important constituent is cocaine.

Medicinal Properties.—The most valuable properties are attributed to coca by the natives of Peru, Bolivia, and some other parts of South America, where it is in enormous use, mixed with lime or wood ashes, for chewing. It is said to possess marvellous sustaining powers, and that its use enables one to undergo great fatigue and want of sleep for a long time in the absence of food. Its use is also said to prevent the difficulty of breathing which is generally experienced in the ascent of long and steep mountains. When employed in excess, however, it is said to produce injurious effects, analogous to those caused by the immoderate consumption of opium and fermented liquors. Its therapeutical properties, as tested by the experiments and practical experience of medical practitioners, prove that it is useful as a nervine tonic and stimulant ; and it is also frequently regarded as a diaphoretic, &c. The powerful and most valuable effects of its alkaloid, cocaine, more especially in the form of the hydrochlorate of cocaine, as a local anæsthetic in operations of the eye, teeth, &c., are, however, undoubted, and have led to its very extensive use recently in this country and elsewhere.

Official Preparations.

Cocainæ Hydrochloras. *Dose.*—$\frac{1}{3}$ to 1 grain. This is also used in the preparation of Lamellæ Cocainæ.

Extractum Cocæ Liquidum. *Dose.*—$\frac{1}{2}$ to 2 fluid drachms.

ORDER 2.—ZYGOPHYLLACEÆ.

1. GUAIACUM OFFICINALE, *Linn.*

Jamaica Guaiacum.   Lignum Vitæ.

(Bentley and Trimen's 'Medicinal Plants,' vol. i. plate 41.)

Habitat.—It is found in most of the West Indian Islands, especially in Jamaica, St. Domingo, and Cuba. It is also found in Columbia and Venezuela on the South American continent.

2. GUAIACUM SANCTUM, *Linn.*

Habitat.—It grows in Cuba, the Bahamas, and St. Domingo, but it is not found in South America. It also occurs in the islet of Key West, south of Florida, and in Southern Florida.

Official Parts or Products and Names.—1. GUAIACI LIGNUM :—the heart-wood of the above two species of Guaiacum. 2. GUAIACI RESINA :—the resin obtained from the stem of the above species, by natural exudation, by incision, or by heat.

1. Guaiaci Lignum.

Guaiacum Wood.

Commerce and Preparation.—Guaiacum wood, which is commonly known in commerce as *Lignum Vitæ*, is chiefly imported from the island of St. Domingo, the best from the city of that name; but some is also derived from Jamaica, the Bahamas, and other parts. It is derived from both the official species, and is usually imported in logs or billets of varying sizes, from which the bark and portions of the alburnum or sap-wood have been removed. But for use in pharmacy, the wood, as thus imported, is officially directed to be deprived entirely of its sap-wood, and the heart-wood reduced to the form of chips, raspings, or shavings.

General Characters and Tests.—The logs are hard, heavy, and tough. A transverse section shows no evident pith ; very indistinct and usually eccentrically arranged rings, and a very marked distinction between the heart-wood and sap-wood. The heart-wood, which forms the principal part of the wood—indeed in some cases the whole—is of various shades of greenish-brown, according to the time it has been exposed, and alone contains resin; while the sap-wood is pale yellow, and contains no resin, and hence, as the activity depends upon the resin, the *heart-wood alone is official.* As seen in the pharmacies, the chips, raspings, or shavings, should be dark greenish-brown, and free from yellowish-white pieces. When chewed for a short time their taste is acrid and somewhat aromatic ; and when rubbed, and more especially when heated, their odour is agreeable and faintly aromatic. When touched with nitric acid, they assume a temporary bluish-green colour ; and if moderately heated in a solution of perchloride of mercury, a bluish-green colour is also produced.

Principal Constituents.—The constituent to which it owes its medicinal properties is the official *resin* (see *Guaiaci Resina*).

2. Guaiaci Resina.

Guaiacum Resin.

Commerce and Extraction.—Guaiacum resin is principally obtained from *Guaiacum officinale.* It is imported mostly from St. Domingo, but also, to some extent, from Jamaica, and elsewhere. It is either a natural exudation ; or is obtained after incisions made into the bark ; or long pieces or billets of the wood, after being much incised in the middle, are suspended horizontally in the air by two upright stakes, then set on fire at the two ends, and the melted resin which runs from the centre is received in a calabash or some other suitable receptacle. The latter is the more general mode adopted for obtaining the resin.

General Characters and Tests.—Guaiacum resin occurs either in tears, or more commonly in masses. The *tears* vary from half an inch to one inch or more in diameter; they are roundish or somewhat oval in form, and are usually covered by a green powder, and in some cases they are more or less agglutinated together into small masses. The ordinary guaiacum of the pharmacies is in large masses, usually containing fragments of wood, bark, and other impurities. These masses are brownish or greenish-brown externally, and when the surface has been rubbed and exposed to air and light, they are covered with a green powder. Guaiacum resin is brittle, breaking with a clean glassy fracture; thin splinters are transparent and greenish-brown; it is easily powdered, the powder being greyish at first, but becoming green by exposure. Its odour, especially when powdered or heated, is somewhat balsamic; and when chewed it leaves an acrid sensation in the throat. It is readily soluble in rectified spirit, the solution producing a clear blue colour when applied to the inner surface of a paring of raw potato; and if paper be wetted with it, and then exposed to the fumes of nitric acid, it also speedily becomes blue.

Principal Constituents.—Two resinous acids, namely, *guaiaconic acid*, about 70 per cent.; and *guaiaretic acid*, about 10 per cent. The former is amorphous, and assumes a blue colour under the influence of oxidising agents, as nitric acid and ferric chloride; the latter is crystalline, and is not coloured blue by oxidising agents. Also about 10 per cent. of a neutral resin termed *guaiac beta-resin*, and small proportions of two crystalline substances, called *guaiacic acid* and *guaiac-yellow*.

Adulterations and Impurities.—It can scarcely be said to be subject to adulteration in this country, although turpentine resin has been stated to be used for that purpose; but, if such be the case, its presence may be readily detected on heating by the terebinthinous odour then evolved. It is, however, often contaminated by a large proportion of impurities, owing to the careless manner in which it is collected.

Medicinal Properties.—Both guaiacum wood and guaiacum resin are stimulant, diaphoretic, and alterative, and probably emmenagogue ; the action of the resin is much more powerful than that of the wood.

Official Preparations.

1. Of GUAIACI LIGNUM:—
   Decoctum Sarzæ Compositum.
2. Of GUAIACI RESINA:—
   Mistura Guaiaci.
   Pilula Hydrargyri Subchloridi Composita.
   Tinctura Guaiaci Ammoniata.   *Dose.*—$\frac{1}{2}$ to 1 fluid drachm.

ORDER 3.—RUTACEÆ.

1. GALIPEA CUSPARIA, *St. Hil.*
Cusparia or Angustura.

(Bentley and Trimen's ' Medicinal Plants,' vol. i. plate 43.)

Habitat.—Eastern parts of Venezuela ; and it appears to extend to New Granada, but has not yet been noticed in Brazil.

Official Part and Name.—CUSPARIÆ CORTEX :—the dried bark.

Cuspariæ Cortex.
Cusparia Bark.

Commerce.—It is imported directly from the northern parts of South America, or indirectly from the West Indies.

General Characters and Test.—This bark, which is commonly known as *Cusparia* or *Angustura Bark*, is in flattish or curved pieces, or in quills which are usually not more than about six inches in length; the bark itself being commonly not more than one-sixth of an inch in thickness, and obliquely cut on its inner edge. It is coated by a yellowish-grey or brownish mottled corky layer, which may be usually scraped off by the nail, the exposed surface then presenting a dark brown resinous appearance; the inner surface is light brown or yellowish-brown, smooth, separable into layers, and occasionally with strips of the wood attached. The fracture is short and resinous, reddish-brown, and exhibiting—more especially when examined by a magnifying lens—numerous white points or lines, from the presence of calcium oxalate. It has a bitter somewhat aromatic taste; and a disagreeable musty odour. The fractured surface touched with nitric acid does not become of an arterial blood-red colour.

Principal Constituents.—About ½ per cent., or somewhat more, of a *volatile oil*, to which its odour is due; and a neutral crystalline principle termed *cusparin*, to which the bitter taste of the bark has been ascribed, but its presence is doubtful. Some chemists have, however, found in cusparia bark a peculiar alkaloid, which they have named *angusturine*.

Substitutions.—The poisonous bark of *Strychnos Nux-vomica* has been substituted for cusparia bark, but the two barks may be readily distinguished as follows :—(1) Nux-vomica bark has a pure and very intense bitter taste, but no odour ; (2) the fractured surfaces of the spurious bark present no white spots or lines; and (3) when touched by concentrated nitric acid, the freshly-fractured surface immediately assumes an arterial blood-red colour, while cusparia bark under the same circumstances becomes dull purplish-red like that of venous blood.

Another false bark has been noticed in the United States and in France ; and its botanical source has been traced by Maisch to *Esenbeckia febrifuga*, Martius. It is known in Brazil as *China Piavi, China du Brésil* (Brazil

Bark), and may readily be distinguished (1) by the dark brown colour of its inner surface; (2) by its fibrous fracture and absence of white marks of calcium oxalate on the fractured surfaces; and (3) by its want of aromatic properties both in taste and odour.

Medicinal Properties.—Stimulant, aromatic, tonic, and perhaps febrifugal in moderate doses; and in large doses it produces nausea and purging.

Official Preparation.—Infusum Cuspariæ.

---

2. RUTA GRAVEOLENS, *Linn.*
Rue.

(Bentley and Trimen's ' Medicinal Plants,' vol. i. plate 44.)

Habitat.—It grows throughout the South of Europe, extending from Spain and Portugal to Greece and the Crimea. It also occurs in the Canaries. In this country it is one of our oldest garden plants, but is not found in a wild state.

Official Product and Name.—OLEUM RUTÆ :—the oil distilled from the fresh herb.

General Characters of the Herb.—When fresh the herb has a green colour, a very strong disagreeable odour, and a bitter acrid nauseous taste. The dried herb is greyish-green, and has a similar taste and odour, but the latter is less powerful.

Principal Constituents.—The properties of the herb are essentially due to the official *volatile oil* described below ; it also contains a *bitter extractive matter*.

Oleum Rutæ.
Oil of Rue.

When fresh the oil is pale yellow or nearly colourless, but it darkens by keeping. It has an acrid bitter taste, and the strong disagreeable characteristic odour of the fresh

herb. Its specific gravity is about o·888 ; it has a neutral
reaction, and is soluble in an equal weight of alcohol.

Medicinal Properties.—Antispasmodic, emmenagogue,
and according to some anthelmintic. In excessive doses it
is an acro-narcotic poison. Externally applied it is rubefacient. It has been regarded from the earliest times as
most useful in warding off contagion, and for preventing
the attacks of noxious insects.

Dose.—1 to 4 minims.

3. BAROSMA BETULINA, *Bart. & Wendl.*
Buchu.

(Bentley and Trimen's ' Medicinal Plants,' vol. i. plate 45.)

4. BAROSMA CRENULATA, *Hook.*

(Bentley and Trimen's ' Medicinal Plants,' vol. i. plate 46.)

5. BAROSMA SERRATIFOLIA, *Willd.*

(Bentley and Trimen's ' Medicinal Plants,' vol. i. plate 47.)

Habitat.—These three species of Barosma are all
natives of the Cape of Good Hope ; but not known elsewhere in a wild state.

Official Part and Name.—BUCHU FOLIA :—the dried
leaves of the above three species of Barosma.

Buchu Folia.
Buchu Leaves.

Commerce.—Buchu leaves are entirely derived from the
Cape of Good Hope, and as imported they consist of the
leaves of one of the above-mentioned species, frequently
more or less mixed with the flowers, fruits, and stalks of
the same.

General Characters.—The leaves of all the species are smooth, coriaceous, serrate, somewhat dentate, or crenate, and marked on the margins, and especially on their under surface, with oil-glands. Their colour is dull yellowish-green above, somewhat paler beneath; they have a strong peculiar penetrating odour; and an aromatic bitterish mint-like taste. Of the three species, those of *Barosma betulina* are the least esteemed, but we have no sufficient evidence of their inferiority. The distinctive characters of the three species may be summed up as follows :—

1. *Barosma betulina.* From a half to three-quarters of an inch long (*fig.* 13, 1), shorter than those of the other

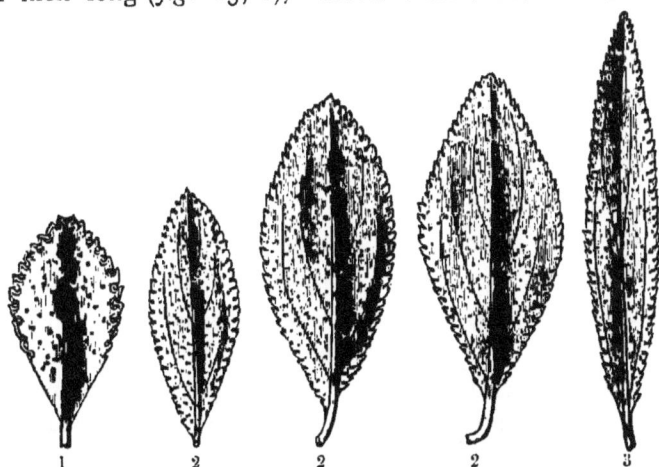

Fig. 13.—Buchu Folia. *Buchu Leaves.* 1. Leaf of *Barosma betulina.* 2, 2, 2. Leaves of *Barosma crenulata.* 3. Leaf of *Barosma serratifolia.* (All enlarged.)

species, and hence sometimes distinguished as *short buchu*; cuneate or rhomboid-obovate, lateral veins indistinct, serrate-dentate, very blunt and usually recurved at the apex; and more cartilaginous in texture than the other species.

2. *Barosma crenulata.* From three-quarters to about an inch and a quarter long (*fig.* 13, 2, 2, 2), thickish, oval-oblong or rhomboid-oval, lateral veins few, curved, somewhat blunt at the apex, narrowed into a distinct petiole at the base, crenate-serrate or finely serrate.

3. *Barosma serratifolia.* From an inch to an inch and a half long (*fig.* 13, 3), and hence, being longer than those of the other species, sometimes termed *long buchu.* They are linear-lanceolate, and have the lower lateral veins distinct, and extending nearly to the apex, giving the leaf a somewhat three-ribbed appearance; equally tapering to base and apex, but the actual apex truncate, and provided with an oil-gland, and sharply and closely serrate at the margins; texture thinner than in the other species.

Adulterations and Substitutions. — As already noticed, the flowers, fruits, and portions of the stalks of the species of *Barosma* are frequently mixed in varying proportions with the official leaves ; and the leaves of other plants have been also offered as buchu ; but the official leaves are readily distinguished by the characters already given. The leaves of an allied plant, *Empleurum serrulatum,* Ait., having, however, been frequently imported and offered for sale for those of *Barosma serratifolia,* require special notice. They are distinguished by being narrower and commonly longer (*fig.* 14), absence of lateral veins, more coarsely serrate margins, distinctly acute apex without an oil-gland, and different odour.

FIG. 14.—Leaf of *Empleurum serrulatum.*

Principal Constituents.—So far as is known, the properties of buchu leaves are essentially due to a *volatile oil* and *mucilage,* but more especially to the former. A peculiar principle, termed *barosmin* or *diosmin,* has been also indicated as a constituent, but nothing definite is known respecting it.

Medicinal Properties.—Buchu is regarded as a stimulant, tonic, diuretic, and diaphoretic.

Official Preparations.
Infusum Buchu. | Tinctura Buchu.

6. PILOCARPUS PENNATIFOLIUS, *Lemaire.*
Pernambuco Jaborandi.

(Bentley and Trimen's 'Medicinal Plants,' vol. i. plate 48.)

Habitat.—It was first found in the southern provinces of Brazil. It also occurs in some of the hotter northern provinces, especially in the neighbourhood of Pernambuco.

Official Part and Name.—JABORANDI :—the dried leaflets.

Jaborandi.
Jaborandi.

Synonym.—Pilocarpi Foliola.

Commerce.—Imported from Brazil, and principally from Pernambuco. As imported, the leaflets are frequently broken up into fragments of various sizes.

General Characters.—When entire the leaflets vary in length, but are commonly three, four, or more inches, and have very short stalks (*fig.* 15). They are oval-oblong or oblong-lanceolate in shape, slightly unequal at the base, obtuse and usually emarginate at the apex, entire and slightly revolute at the margins, and coriaceous in texture. Their upper surface is glabrous, except when young, and dull green ; the under surface is paler, often somewhat hairy, with a very prominent midrib, and marked irregularly all over with minute glands, which give a pellucid dotted appearance to the leaflets when held against the light. Odour when bruised slightly aromatic ; taste on chewing slightly bitter and aromatic at first, but subsequently producing a warm tingling sensation in the mouth and an abundant flow of saliva.

Principal Constituents.—The activity of Jaborandi as a medicinal agent is essentially due to a peculiar alkaloid named *pilocarpine.* This is a colourless, soft, odourless, amorphous mass, which forms crystallised salts with acids,

G

and one of which, the nitrate, is official. Jaborandi also
contains a little *volatile oil.*

Medicinal   Properties. —
Jaborandi is a powerful diaphoretic and sialagogue. The
official Nitrate of Pilocarpine
also possesses similar properties in a marked degree ; it
may be taken by the mouth
or used hypodermically ; it is
antagonistic to atropine, and
an   antidote   therefore   to
poisoning by that alkaloid.

Official Preparations.

Extractum Jaborandi.
Infusum Jaborandi.
Pilocarpinæ Nitras. *Dose.*
—$\frac{1}{20}$ to $\frac{1}{2}$ grain.
Tinctura Jaborandi.

Other Kinds of Jaborandi.

(*Not Official.*)

The leaflets of *Pilocarpus
Selloanus,* Engler, which is
probably only a variety of *P.
pennatifolius,* have also been
imported, under the name of
Jaborandi, from Rio Janeiro.
They are readily distinguish-

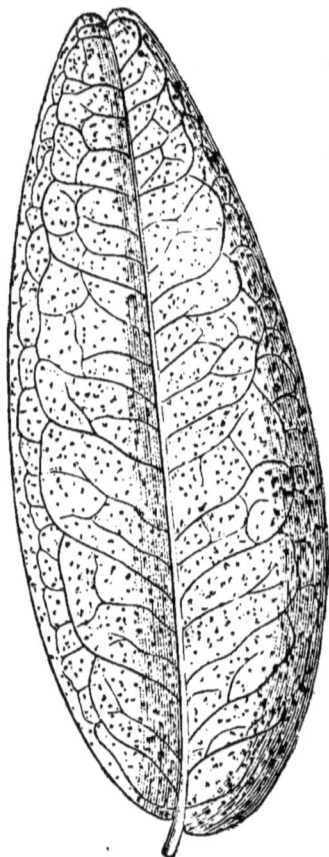

Fig. 15.—Under surface of a leaflet of
Pilocarpus pennatifolius. (After
Holmes.)

able, if attention be given to the characters of the official
Jaborandi.   They are regarded as less active.   Several
Piperaceous plants, and other plants of the order Rutaceæ,
are also known under the name of Jaborandi in Brazil, but
do not appear to be exported.

7. CITRUS VULGARIS, *Risso.*

Bitter Orange. Seville Orange.

Synonym.—CITRUS BIGARADIA, *Duhamel.*

(Bentley and Trimen's ' Medicinal Plants,' vol. i. plate 50.)

Habitat.—In Northern India a wild orange tree is found which is the supposed parent of the cultivated orange, whether Sweet or Bitter. The Bitter Orange is extensively grown in the warmer parts of the Mediterranean district, especially in Spain.

Official Parts or Products and Names.—1. AURANTII FRUCTUS :—the ripe fruit. 2. AURANTII CORTEX :—the dried outer part of the rind or pericarp. 3. AQUA AURANTII FLORIS :—the distilled water of the flowers of the Bitter Orange tree, Citrus vulgaris, *Risso* ; and of the Sweet Orange tree, Citrus Aurantium, *Risso.*

1. Aurantii Fructus.
Bitter Orange.

Commerce.—It is imported from the South of Europe, and is commonly known as the Bitter Orange or Seville Orange.

General Characters.—Fruit globular, except at the two ends, where it is somewhat compressed ; about the size of the sweet orange, but the pericarp is rougher, darker in colour, being deep orange-red or red, the pulp very bitter and sour, and the rind more aromatic, and very bitter.

This fruit is simply introduced into the British Pharmacopœia as the source of fresh bitter orange peel.

2. Aurantii Cortex.
Bitter Orange Peel.

Synonym.—Aurantii Pericarpium.

Preparation and Commerce.—The peel is removed in one long spiral piece by cutting the rind with a sharp knife, and taking care not to include more than can be helped of the inner white spongy portion. It is then dried. The peel dried in Britain is preferred, but it is also largely imported from Malta and other parts in the South of Europe.

General Characters.—In thin flattish pieces, or in curled bands or strips, somewhat uneven, glandular, and of a deep orange-red colour externally, and white within from a portion of the inner spongy layer of the rind not having been removed. It has a pleasant aromatic odour, and an aromatic bitter taste.

Principal Constituents.—The properties of bitter orange peel are essentially due to a *volatile oil*, and a bitter principle termed *hesperidin*. The volatile oil is known in commerce as *Essence de Bigarade.*

Medicinal Properties.—Bitter orange peel is a bitter aromatic tonic and stimulant; it is a pleasant addition to other bitters; and its preparations are useful vehicles for the exhibition of other medicines.

Official Preparations.

1. Of AURANTII FRUCTUS :—
    Tinctura Aurantii Recentis.

2. Of AURANTII CORTEX:—

Infusum Aurantii.
Infusum Aurantii Compositum.
Infusum Gentianæ Compositum.

Spiritus Armoraciæ Compositus.
Tinctura Aurantii.
Tinctura Cinchonæ Composita.

Tinctura Gentianæ Composita.

Tinctura Aurantii is also used in the preparation of Mistura Ferri Aromatica, Syrupus Aurantii, and Tinctura Quininæ.

3. Aqua Aurantii Floris.

Orange Flower Water.

This is described under 'Citrus Aurantium.

8. CITRUS AURANTIUM, *Risso.*

Sweet Orange.

(Bentley and Trimen's ' Medicinal Plants,' vol. i. plate 51.)

Habitat.—Probably a native of Northern India (see Citrus vulgaris). It is very abundantly cultivated in many parts of the Mediterranean district, and in Portugal, &c.

Official Product and Name.—AQUA AURANTII FLORIS :— the distilled water of the flowers of the Bitter Orange tree, (Citrus vulgaris, *Risso*) ; and of the Sweet Orange tree (Citrus Aurantium, *Risso*).

The Orange Flower Water of commerce is usually three times the strength of that employed in former years.

Aqua Aurantii Floris.

Orange Flower Water.

Production and Commerce.—This is commonly prepared at Cannes, Nice, and Grasse, in the South of France, and more generally from the flowers of the Bitter Orange tree, as these yield the more fragrant product. When the fresh flowers are distilled with water, Orange Flower Water passes over together with Oil of Neroli, which floats on the surface, and when this is removed we have the Orange Flower Water of commerce.

General Characters and Test.—Orange flower water is nearly colourless, or with a slight greenish-yellow tint. It has a very fragrant odour and a bitter taste. From being distilled, or kept, in copper or lead vessels, it may contain metallic impurity. Hence the test of the British Pharmacopœia :—' Not coloured by sulphuretted hydrogen.'

Composition.- The characters of orange flower water are especially due to the *volatile oil*, which is known in commerce as Oil of Neroli.

Medicinal Properties.—Orange flower water is a slight

nervine stimulant; but in this country it is only used in medicine on account of its agreeable odour and as a flavouring agent.

Official Preparation.—Syrupus Aurantii Floris.

---

9. CITRUS LIMONUM, *Risso.*

Lemon.

(Bentley and Trimen's ' Medicinal Plants,' vol. i. plate 54.)

Habitat.—This tree is a native of the forests of Northern India. It is now largely cultivated throughout the Mediterranean region, and in Spain, Portugal, and all tropical and subtropical countries.

Official Parts or Products and Names.—1. Limonis Cortex :—the outer part of the rind or pericarp of the fresh fruit. 2. Oleum Limonis :—a volatile oil obtained by mechanical means from fresh lemon peel. 3. Limonis Succus:—the freshly expressed juice of the ripe fruit.

1. Limonis Cortex.

Lemon Peel.

Synonym.—Limonis Pericarpium.

Commerce.—Lemons (from which the peel and juice are obtained) are imported from Southern Europe, more especially from Sicily, but also to some extent from Spain, and elsewhere.

General Characters.—Fresh lemon peel should be in thin flattish or more or less curved pieces or bands, of a pale yellow colour, somewhat rough on the outer surface in consequence of the oil glands imbedded in the tissue beneath, and with very little of the white spongy portion of the rind on its inner surface. It has a strong peculiar fragrant odour; and a warm aromatic bitterish taste.

Principal Constituents.— *Volatile oil* (see *Oleum Limonis*), and a bitter principle termed *hesperidin.*

Medicinal Properties.— Lemon peel is an aromatic stomachic. It is especially employed as a flavouring agent

2. Oleum Limonis.

Oil of Lemon.

Production and Commerce.—The best, and nearly all the oil of lemon which reaches this country is prepared in Sicily and Calabria by a mode of expression known as the *sponge process* (see 'Pharmacographia,' 2nd ed. page 118). It is imported from Palermo and Messina. Some is also obtained in France, at Nice and Mentone, either by subjecting the grated peel to distillation with water, or by what is known as the *écuelle à piquer* process (see 'Pharmacographia,' 2nd ed. page 119).

General Characters.—Oil of lemon, when prepared by the *sponge* or *écuelle à piquer* process, is a pale yellow limpid liquid, with a very fragrant odour, and a warm bitterish aromatic taste. The distilled oil is colourless and far less fragrant. Both oils are dextrogyre.

Adulteration.—Nearly all the oil of lemon of commerce is more or less diluted with oil of turpentine, or with the cheaper distilled oil.

Medicinal Properties. — Internally administered, it is stimulant and carminative; and when applied externally, stimulant and rubefacient. Its use in medicine, however, is essentially confined to giving an agreeable odour and flavour to other medicines.

3. Limonis Succus.

Lemon Juice.

Commerce.—Concentrated lemon juice is imported in enormous quantities from Messina and elsewhere for citric acid manufacturers; but for use in medicine, it should be pressed, when wanted, from the ripe fruit, as directed in the British Pharmacopœia.

General Characters.—A slightly turbid yellowish liquid, with a sharp acid taste. It is odourless in itself, but it has usually a grateful lemon-like odour from the presence of some oil of lemon expressed from the rind. Its specific gravity varies from 1·035 to 1·045.

Principal Constituents.—*Citric acid* is the essential constituent ; but the amount varies from 36 to 46 grains in one fluid ounce. Lemon juice also contains *mucilage, sugar, inorganic salts*, and other substances.

Medicinal Properties.—Refrigerant and antiscorbutic.

Official Preparations.

1. Of LIMONIS CORTEX :—

Infusum Aurantii Compositum.

Infusum Gentianæ Compositum.

Oleum Limonis. (*See page* 87.)

Syrupus Limonis.

Tinctura Limonis.

2. Of OLEUM LIMONIS :—
   Linimentum Potassii Iodidi cum Sapone.
   Spiritus Ammoniæ Aromaticus.

3. Of LIMONIS SUCCUS :—
   Syrupus Limonis, which is also used in the preparation of Liquor Magnesii Citratis.

4. Other Official Preparations of the Lemon.

Besides the official parts, products, and preparations from the Lemon already alluded to, Citric Acid is also official in the British Pharmacopœia. It is described as 'an acid prepared from lemon juice, or from the juice of the fruit of Citrus Bergamia, *Risso & Poit.* (Citrus Limetta, *DC.*), the Lime.' There are also several official citrates in the British Pharmacopœia.

The description of citric acid, however, further than indicating its source, is not within our province.

10. ÆGLE MARMELOS, *Correa.*

Bael. Bengal Quince.

(Bentley and Trimen's ' Medicinal Plants,' vol. i. plate 55.)

Habitat.—The Bael tree is found throughout the Indian Peninsula. It is extensively cultivated, and frequently planted near the Hindoo temples.

Official Part and Name.—BELÆ FRUCTUS:—the dried half-ripe fruit.

Belæ Fructus.

Bael Fruit.

Collection and Commerce.—The fruit is essentially collected from cultivated plants. It is imported from Malabar and Coromandel. It should be obtained in a half-ripe state, and carefully dried ; but the ripe fruit, as shown by the fully-matured seeds, is sometimes seen in commerce.

General Characters.—Roundish, and about the size of a large orange when entire, and covered with a hard woody nearly smooth rind ; but it is usually imported in dried more or less twisted slices, or in fragments consisting of portions of the rind and adherent dried pulp and immature seeds. The *rind* is about one-eighth of an inch thick, hard, and covered with a nearly smooth pale-brown or greyish firmly adherent epicarp. The *pulp* is firm and brittle, and of an orange-brown or cherry-red colour externally ; but when broken, it is seen to be nearly colourless within. It has no odour, and its taste is simply mucilaginous and very slightly acid.

Principal Constituents.—*Mucilage* ; and probably a very little *tannic acid.*

Substitutions.—Several fruits, or portions of fruits, have been substituted for Bael Fruit : such as the Wood Apple, from *Feronia Elephantum*, Correa ; Pomegranate rind, &c. ; but these are readily distinguished by the characters given

above.   The more common substitution is, as first described by the author, the dried rind of the Mangosteen fruit
(*Garcinia Mangostana*).   This is known from bael fruit by
its darker colour, greater thickness, freedom internally from
adherent pulp, easily separable epicarp, and by some of the
pieces having dark-coloured wedge-shaped radiating projecting stigmas upon them.

Medicinal Properties.—It is commonly regarded as a
mild astringent, and useful in diarrhœa and dysentery.

Official Preparation.—Extractum Belæ Liquidum.

ORDER 4.—SIMARUBACEÆ.

1. SIMARUBA AMARA, *Aublet.*
Simaruba.

(Bentley and Trimen's 'Medicinal Plants,' vol. i. plate 56.)

Habitat.—In Northern Brazil, Guiana, and the islands
of Dominica and St. Vincent.   A closely allied species, *S.
glauca*, DC., occurs in Jamaica and in the West Indies
generally, and in South Florida and Central America; this
has been often confounded with *S. amara.*

Part Used and Name.—SIMARUBÆ RADICIS CORTEX :—
the dried bark of the root of Simaruba amara and of S.
glauca.

(*Not Official.*)

Simarubæ Radicis Cortex.

Simaruba Bark.

Commerce.—It is usually imported from Jamaica, and is
therefore generally derived from *S. glauca*, as *S. amara* is
not found in that island.   (See *Habitat.*)

General Characters.—In flattened longitudinally-folded

pieces, several feet in length, and from one to three inches or more in breadth ; or in transversely-cut portions of these a few inches only in length; or in more or less evident quills. Its outer surface, from which the corky layer is often more or less removed, is greyish, greyish-brown, or whitish-yellow in colour ; somewhat rough, warty, and marked with irregular transverse ridges. Beneath the corky layer it is darker coloured; and internally, pale yellowish-white, or yellow when fresh. It is very fibrous, extremely tough and difficult to powder, and light in weight; inodorous, but with a very purely bitter taste.

Principal Constituents.—Its activity is said to arise essentially from *quassin*, the same principle to which the properties of the official quassia wood are due.

Medicinal Properties.—Tonic and febrifugal.

Preparation.

The best form of administration is the Infusion, which may be made by macerating half an ounce of the bruised bark, for two hours, in a pint of boiling water, and then straining.

---

2. PICRÆNA EXCELSA, *Lindl.*

Jamaica Quassia. Bitter Wood. Bitter Ash.

Synonym.—QUASSIA EXCELSA, *Swartz.*

(Bentley and Trimen's ' Medicinal Plants,' vol. i. plate 57.)

Habitat.—Common in Jamaica, and also found in some other West Indian Islands, as St. Kitts, Antigua, and St. Vincent.

Official Part and Name.—QUASSIÆ LIGNUM:—the chips, shavings, or raspings of the wood.

Quassiæ Lignum.

Quassia Wood.

Commerce and Preparation.—It is imported from Jamaica in billets or logs of varying lengths and sizes, but frequently as thick as a man's thigh. The wood is dense, tough, porous, and of a pale yellowish-white colour. It is covered by a dark-grey or brownish bark, which is commonly stripped off; and the wood, in the form of shavings, chips, or raspings, is alone met with in the pharmacies.

General Characters and Test.—The shavings, chips, or raspings have a very pale yellowish-white colour; somewhat silky appearance; and have no odour, but an intense and purely bitter taste. An infusion does not become black or bluish-black on the addition of a persalt of iron.

Principal Constituents.—Its essential constituent is a neutral crystallisable bitter principle, termed *quassin* or *quassiin*, the same principle which is said to exist in simaruba bark. It contains *no tannic acid* or other astringent matters.

Adulterations and Substitutions.—The chips, shavings, and raspings of other woods may be sometimes found mixed with, or substituted for, those of quassia. The pure and intensely bitter taste of the genuine drug, combined with the fact that, as quassia wood contains no tannic acid, its infusion is not coloured of a black or bluish-black hue by a persalt of iron, are generally ready means of distinguishing it.

Medicinal Properties.—Tonic and stomachic. It is also sometimes regarded as anthelmintic, febrifugal, and very slightly narcotic.

Official Preparations.

Extractum Quassiæ. | Infusum Quassiæ.
Tinctura Quassiæ.

ORDER 5.—BURSERACEÆ OR AMYRIDACEÆ.

1. BALSAMODENDRON MYRRHA, *Nees.*

Myrrh.

(Bentley and Trimen's ' Medicinal Plants,' vol. i. plate 60.)

Habitat.—First seen by Ehrenberg at Ghizan or Gison, on the coast of the Red Sea in Southern Arabia, in 1820-25; but not since noticed there. It was found, however, by Hildebrandt in 1873, and by Wykeham Perry in 1878, in Somali Land, Eastern Africa.

Official Product and Name.—MYRRHA :—a gum-resinous exudation obtained from the stem

Myrrha.

Myrrh.

Collection and Commerce.—Myrrh exudes spontaneously from the stem of the myrrh plant, and, so far as is known, it is chiefly, if not entirely, collected in the Somali Country and some other parts of Eastern Africa; from whence it is brought to the great fair of Berbera, where it is purchased by the Banians of India, and thence chiefly forwarded to Bombay by way of Aden. At Bombay it is separated from bdellium and other foreign substances, and sorted into myrrh of different qualities, and then exported to Europe and elsewhere. Some myrrh is also shipped directly from Aden to this country.

General Characters. Myrrh is in roundish or irregular-formed tears, or in masses of agglutinated tears, varying in size from small grains up to pieces as large as a hen's egg, or even that of a man's fist. It is opaque, of a reddish-brown or reddish-yellow colour externally, dry, and more or less covered by a fine powder. It is brittle, and when broken, the fractured surface presents an irregular, somewhat translucent, rich brown, oily appearance ; and is frequently marked with opaque whitish spaces or striæ, which the ancients compared to the light-coloured marks at the base of the finger-nails. The odour is agreeable, aromatic, and peculiar; and the taste bitter, aromatic, and acrid.

Principal Constituents.—Myrrh is composed in varying proportions of *gummy matters, resin,* and *volatile oil.* The odour of myrrh is due to the volatile oil; but its more essential properties reside in the resin, in which the *bitter principle of myrrh* is contained. This bitter principle is regarded by Flückiger as a glucoside.

Medicinal Properties.—Stimulant, tonic, expectorant; it is also regarded as antispasmodic and emmenagogue. Locally applied it is stimulant.

Official Preparations.

Decoctum Aloes Compositum.	Pilula Asafœtidæ Composita.
Mistura Ferri Composita.	Pilula Rhei Composita.
Pilula Aloes et Myrrhæ.	Tinctura Myrrhæ.

Adulterations and other Kinds of Myrrh.

The inferior qualities of myrrh are frequently more or less adulterated with various gums, resins, and other analogous substances. These are readily distinguished from the official or true myrrh, either by the transparency of their fractured surfaces, or, in some cases, by their greater opacity; and always by their differences of odour, taste, and other characters. There are also two other kinds of myrrh, which have been distinguished by pharmacologists under the names of *Arabian Myrrh* and *East India Myrrh,* both of which are frequently sold by the dealers as true myrrh.

1. *Arabian Myrrh.*—This appears to be obtained from a different but allied species of *Balsamodendron* to that yielding the official or true myrrh. It is collected in Southern Arabia to the eastward of Aden. Arabian myrrh is in irregular gummy-looking masses, rarely exceeding one and

a half inch in length. It is brittle like true myrrh, but its
fractured surfaces are less oily-looking than it, and the
whitish opaque markings are absent. Its taste and odour
resemble the official myrrh.

2. *East India Myrrh, Bissa Bol*, or *Habaghadi.*—This,
like true myrrh, is collected in the Somali Country, but its
botanical source is unknown, although probably a species of
Balsamodendron. The finer qualities of this myrrh have
a considerable resemblance to the official myrrh, but the
colour is darker. Its chief distinctive characters are, its
peculiar odour, which is very different from true myrrh, and
its strong almost acrid taste. It is usually a very impure
substance.

2. CANARIUM COMMUNE, *Linn.*
Manila Elemi.

(Bentley and Trimen's ' Medicinal Plants,' vol. i. plate 61.)

Habitat.—Native of Amboyna, Luzon, Sunda, the
Moluccas, and Penang.

Official Product and Name.—ELEMI :—a concrete resinous exudation.

Elemi.
Manila Elemi.

Botanical Source and Commerce.—Although Elemi is
placed under the above plant, its botanical source, as stated
in the British Pharmacopœia, 'is undetermined, but is
sometimes referred to Canarium commune, *Linn.*' Indeed,
there is no reliable evidence that such is its source. It is
chiefly, if not entirely, imported from Manila, hence its
common name.

General Characters.—When quite pure and fresh, it is
soft, granular, resinous, and colourless, but by keeping it
becomes harder, somewhat friable, and of a pale yellow tint.

As seen in commerce, however, it is frequently mixed with chips of wood and other impurities, and sometimes has a greyish or blackish colour. Its odour is strong and fragrant, somewhat terebinthinous, but also resembling fennel and lemon ; and it has a bitterish disagreeable pungent taste. When moistened with rectified spirit it breaks up into small particles, which, when examined by the microscope, are seen to consist in part of acicular crystals.

Principal Constituents.—Manila Elemi is an oleo-resin, the *volatile oil* being in the proportion of about 10 per cent. The other constituents which have been described, some of which, however, require further examination, are about 60 per cent. of *amorphous resin*, soluble in cold rectified spirit; from 20 to 25 per cent. of a neutral crystallisable resin, soluble in hot rectified spirit, termed *amyrin* ; and three other crystalline principles in small proportion, namely, *bryoidin*, *breïdin*, and *elemic acid.*

Medicinal Properties.—It has stimulant properties like those of the turpentines ; but it is only used externally.

Official Preparation.—Unguentum Elemi.

ORDER 6.—RHAMNACEÆ.

1. RHAMNUS CATHARTICUS, *Linn.*
Buckthorn.

(Bentley and Trimen's ' Medicinal Plants,' vol. i. plate 64.)

Habitat.—It is indigenous throughout the greater part of Europe, except the extreme north, and extends eastwards into Siberia. It is also found in Algeria. It is not unfrequent in the South of England, but becomes rare northwards, and is not a native of Scotland.

Product Used and Name.—RHAMNI SUCCUS :—the recently expressed juice of the ripe fruits.

(Not Official.)

Rhamni Succus.

Buckthorn Juice.

Collection.—Buckthorn fruits, commonly, but incorrectly, called berries, are more especially collected in Hertfordshire, Oxfordshire, Buckinghamshire, and Wiltshire ; and when the juice is required, it is better not to obtain this from the collectors, as it is often largely diluted with water, but to press it direct from the fresh fruit, which is generally ripe about the middle of September.

General Characters of the Fruit.—The fresh ripe fruits are about the size of a pea, smooth, round, fleshy, black and shining, umbilicated at the apex by the remains of the style, and surrounded at the base by the persistent portion of the calyx-tube in the form of a circular disc, and supported by a slender stalk. The epicarp is thin, and envelopes a scanty pulp, in which are contained four hard indehiscent dark-brown stones or pyrenes, each of which contains a solitary seed. The presence of four pyrenes at once distinguishes Buckthorn fruits from those of the other British species, *Rhamnus Frangula*, Linn., Alder Buckthorn, which have only two. One or two of the stones are, however, frequently abortive.

General Characters of the Juice.—The fresh juice is green, has an acid reaction, a very disagreeable odour, and a taste which, although sweetish at first, is ultimately disagreeably bitter and somewhat acrid. Its specific gravity is from 1·070 to 1·080. By keeping, its specific gravity is reduced to 1·035, and its colour changed to red.

Principal Constituents.—Nothing definite is known of the purgative principle or principles of buckthorn juice ; but two substances have been more especially indicated as constituents, namely, *rhamnocathartin* and *rhamnin*: the first a yellowish bitter uncrystallisable substance, and the latter a yellow crystalline glucoside.

H

Medicinal Properties.—Powerful hydragogue cathartic, but almost out of use at the present day, except in veterinary practice.

Preparation.

It is commonly used in the form of a Syrup, a formula for which may be found in the British Pharmacopœia of 1867.

---

2. RHAMNUS FRANGULA, *Linn.*

Alder Buckthorn. Black Alder.

(Bentley and Trimen's ' Medicinal Plants,' vol. i. plate 65.)

Habitat.—Frequent in England, but very rare in Scotland. It is also found throughout Europe, and extending into Siberia, the Caucasus, and the Mediterranean coast of Africa.

Official Part and Name.—RHAMNI FRANGULÆ CORTEX: the dried bark.

Rhamni Frangulæ Cortex.

Frangula Bark.

Synonym.—Cortex Frangulæ.

Collection.—It is officially directed to be ' collected from the young trunk and moderate-sized branches, and kept at least one year before being used.'

General Characters.—In small quills, the bark itself being about one-twenty-fifth of an inch or somewhat more in thickness, and covered with a greyish-brown or blackish-brown corky layer, which is marked with numerous transversely-lengthened whitish lenticels ; the inner surface is smooth, and of a bright reddish-yellow or brownish-yellow colour. Fracture short and purplish externally, but somewhat fibrous and yellowish within. It has no very marked odour ; but a pleasant sweetish slightly bitter taste.

Principal Constituents.—A yellow crystallisable principle termed *rhamnoxanthin* or *frangulin*, an amorphous *bitter principle* of a resinous nature, *emodin*, *resin*, and a little *tannic acid*, &c. The active principle is not determined, but the bitter amorphous principle is said to have a purgative action.

Medicinal Properties.—The fresh bark is stated to be emetic ; but the dried bark, which has been kept at least one year, is alone official. It acts as a pleasant laxative, and is also reputed to be tonic and diuretic.

Official Preparations.

Extractum Rhamni Frangulæ.
Extractum Rhamni Frangulæ Liquidum.

---

3. RHAMNUS PURSHIANUS, *DC.*

Sacred Bark.

(Hooker's Flora Boreali-americana, plate 43.)

Habitat.—It is found on the west coast of North America, from the British possessions southward to California.

Official Part and Name.—RHAMNI PURSHIANI CORTEX: the dried bark.

Rhamni Purshiani Cortex.

Sacred Bark.

Synonym.—Cascara Sagrada.

Collection and Commerce.—It is collected in the Rocky Mountains and westward to the Pacific. It is imported from the United States either in its entire state ; or in flattened packets consisting of small pieces of the bark compressed into a more or less compact mass. In commerce it is commonly known as *Sacred Bark* or *Chittem Bark*.

General Characters.—When imported in its entire state,

it is in quills or more or less incurved pieces of varying lengths and sizes, the bark itself being from about one-twenty-fifth to one-eighth of an inch thick, or more. Externally, it is smooth or nearly so, and covered with a greyish-white layer, which is frequently marked with spots or patches of adherent lichens, and is usually easily removed. Beneath the surface it is violet-brown, reddish-brown, or brownish ; and internally, yellowish-brown or dull purplish- or reddish-brown, and nearly smooth, although somewhat striated longitudinally and wrinkled. Its fracture is short, except internally, where it is slightly fibrous, more especially in the larger pieces. It has little or no marked odour after exposure to the air for some time, but when enclosed in a bottle or otherwise it has a very peculiar odour, which has been compared to that of a byre ; it has a bitter taste.

Principal Constituents.—*Three resins,* a *white sublimable crystalline principle, tannic acid,* a small quantity of *volatile oil* to which its peculiar odour is due; and other unimportant constituents.

Medicinal Properties.—In large doses cathartic, and in small doses tonic and stomachic.

Official Preparations.

Extractum Cascaræ Sagradæ.
Extractum Cascaræ Sagradæ Liquidum.

———   ———

Order 7.—VITACEÆ.

VITIS VINIFERA, *Linn.*
Grape Vine.

(Bentley and Trimen's ' Medicinal Plants,' vol. i. plate 66.)

Habitat.—The Grape Vine under its very numerous varieties has been cultivated from the earliest times, but it appears to be indigenous to the Caucasian provinces of Russia, that is, between the eastern end of the Black Sea and the south-western shores of the Caspian ; and extending southward into Armenia. It is cultivated in nearly all the warmer and drier parts of the temperate regions of both hemispheres.

Official Part and Name.—Uvæ :—the dried ripe fruit. Imported from Spain.

Uvæ.

Raisins.

Synonym.—Uvæ Passæ.

Preparation and Commerce.—The fresh fruits, known as Grapes, are either dried by the heat of the sun, or partly by the sun's heat and partly by artificial heat. They are imported in enormous quantities from Spain and Asia Minor, and the variety known as *Currants*, which is a corruption of Corinth, from the Ionian Islands. The raisins imported from Spain are alone official, as already mentioned.

General Characters.—Raisins are more or less shrivelled in appearance, smooth, compressed, free from sugary or saline incrustation ; agreeably fragrant, and with a soft very sweet pulp.

Varieties.—There are several commercial varieties of Raisins, the more important of which are Muscatel, Valencia, Sultanas, and Currants, the latter being distinguished formerly in the old pharmacopœias by the name of *Uvæ passæ minores*. *Muscatel Raisins*, which are imported from Malaga, are the finest. They are in entire bunches, and are carefully dried and packed for the market in boxes. *Valencia Raisins*, which derive their name from their place of growth in Spain, are separated from their stalks. They are the variety commonly employed in pharmacy. *Sultana Raisins*, like the Valencia Raisins, have been separated from their stalks, and are stoneless, or without seeds. They are imported from Smyrna. The small variety known as *Corinthian*

Raisins or *Currants,* which are obtained from the Ionian Islands, are also without seeds, and separated from their stalks.

Principal Constituents.—The skin and seeds contain *tannic acid* ; the pulp, *grape sugar, acid tartrate of potash, gum, malic acid,* &c.

Medicinal Properties.—Raisins are demulcent and slightly refrigerant. They are used in medicine simply for flavouring.

Official Preparations.

Tinctura Cardamomi Composita.

Tinctura Sennæ, which is also used in the preparation of Mistura Sennæ Composita.

Tinctura Cardamomi Composita is also used in four other official preparations.

Order 8.—ANACARDIACEÆ.

1. PISTACIA LENTISCUS, *Linn.*

Mastich Tree. Lentisk.

(Bentley and Trimen's ' Medicinal Plants,' vol. i. plate 68.)

Habitat.—It is a native of the Mediterranean region, from Spain to Syria, and has also been found in Portugal, the Canary Islands, and in Somali Land. It is cultivated in the northern part of the island of Scio.

Official Product and Name.—MASTICHE :—a concrete resinous exudation obtained by making incisions in the bark of the stem and large branches.

Mastiche.
Mastich.

Collection and Commerce.—In the island of Scio, where Mastich is exclusively collected, and from male plants only, it is obtained as follows : About the middle of June vertical incisions are made very close together in the bark of the stem and large branches, and in a short time afterwards the resin exudes and soon hardens and dries, commonly on the bark in separate tears, or some drops to, or flows towards, the ground, where flat pieces of stone are placed to preserve these droppings from dirt. This latter mastich is usually in small irregular masses.

General Characters.—The best mastich is in small rounded, irregular, oblong, or pear-shaped tears, which have a pale yellow colour, and are either opaque and dusty on their surface, or, generally at the present time, have a glassy and transparent appearance. Mastich is brittle, and breaks with a conchoidal vitreous fracture. It has an agreeable somewhat balsamic and terebinthinous odour; and when chewed it becomes soft and plastic, and has a mild resinous taste. Its specific gravity is about 1·065, and it melts about 226°. It is entirely soluble in ether.

Inferior mastich is in separate tears, or in masses of irregular form. It is darker coloured, less transparent, and frequently contaminated with earthy impurities, pieces of bark, &c.

Principal Constituents.—Mastich contains a very minute proportion of *volatile oil*, to which its odour is due, and two resins, one forming about 90 per cent. and having acid properties, hence termed *masticic acid*, which is soluble in alcohol and ether ; and the other in the proportion of about 10 per cent., called *masticin*, insoluble in alcohol, but soluble in ether. Masticic acid is also known as *alpha-resin of mastich*, and masticin as *beta-resin of mastich*.

Medicinal Properties.— Mildly stimulant and diuretic like the ordinary coniferous turpentines, but in this country its use as an internal remedy has become obsolete. Mastich dissolved in alcohol or ether forms, however, a useful application to a carious tooth to relieve toothache ; and it is also used as a temporary stopping for carious teeth. But the chief employment of mastich is, at Constantinople and in the East, as a masticatory.

2. PISTACIA TEREBINTHUS, *Linn.*

Chian Turpentine.

(Bentley and Trimen's ' Medicinal Plants,' vol. i. plate 69.)

Habitat.—It is a common wild plant throughout the islands and shores of the Mediterranean, and in Asia Minor; and following Hanbury, and including in it *P. atlantica*, Desf., *P. palæstina*, Boiss., and *P. cabulica*, Stocks, it is found in different forms in the Canaries, Northern Africa, Syria, and eastward to Afghanistan.

Product Used and Name.—TEREBINTHINA CHIA :—a liquid oleo-resin obtained by making incisions in the bark of the stem and large branches.

(*Not Official.*)

Terebinthina Chia.

Chian or Cyprian Turpentine.

Collection and Commerce.—Chian turpentine, like mastich, is exclusively derived from the island of Scio, where it exudes to some extent spontaneously; but it is almost entirely obtained by incising the bark of the stem and large branches in the spring of the year, and the exuded juice which hardens by exposure during the night is afterwards scraped from the parts down which it has flowed, and from the flat stones placed at the base of the stem to receive it. It is then liquefied by exposure to the heat of the sun, and strained to separate it from earthy and other extraneous substances.

General Characters. — Chian turpentine, when freshly prepared, is a transparent very thick tenacious liquid, of a greenish-yellow or yellowish-brown colour; but as seen in commerce, it is either a soft solid, or, from exposure to the air, hard and brittle. When solid it is opaque in mass, and of a dull brown colour; but when heated and pressed between slips of glass, if pure, it should be transparent and greenish-yellow. It has an agreeable odour, faintly terebinthinous, but sometimes described as between that of fennel and mastich ; it has little taste, but slightly resembles that of mastich. When old and dry Chian turpentine is covered by a whitish powder, and this when examined under the microscope shows no trace of crystalline structure.

Adulterations.—Chian turpentine is rarely pure, but commonly adulterated with the ordinary coniferous turpentines, or the latter are substituted for it. The characters given above, if carefully noted, will readily distinguish the true from the spurious drug.

Principal Constituents.—When freshly prepared it contains about $14\frac{1}{2}$ per cent. of a *volatile oil* having the odour of the drug ; and the remaining constituent is *a resin*, which from its complete solubility in alcohol is usually regarded as identical with the *alpha-resin* of mastich.

Medicinal Properties.—Stimulant and diuretic like the coniferous turpentines ; but as a medicine it had become obsolete until within the last few years, when on account of its reputed value in cancer, more especially in that of the uterus, it was again largely employed ; but, as no marked benefit was obtained, it has again gone out of use.

SERIES III.—*CALYCIFLORÆ.*

ORDER 1.—LEGUMINOSÆ.

1. CYTISUS SCOPARIUS, *Link.*
Broom.

Synonym.—SAROTHAMNUS SCOPARIUS, *Koch.*

(Bentley and Trimen's 'Medicinal Plants,' vol. ii. plate 70.)

Habitat.—This is a common plant in this country and throughout Western Europe ; but it becomes rare in the Central and Eastern parts, and in the region of the Mediterranean.   It occurs, however, in Italy, and in Central and Northern Russia, extending even to Siberia.

Official Part and Name.—SCOPARII CACUMINA : — the fresh and dried tops.   From indigenous plants.

Scoparii Cacumina.

Broom Tops.

General Characters. — Straight, wand-like, flexible, branched, with five wing-like angles, yellowish-green or dark green, nearly smooth, tough.   Leaves usually absent, or, when present, without stipules, alternate ; small, sessile and simple above, stalked and trifoliate below.   Taste bitter and nauseous ; odour, when fresh and bruised, peculiar, but this is nearly lost in drying.

Principal Constituents.—A volatile alkaloid, which is an oily and colourless liquid named *sparteine*, and a neutral or somewhat acid principle termed *scoparin.*   The former possesses narcotic properties, and is very poisonous; and the latter is generally regarded as the diuretic principle of broom tops, and is non-poisonous.

Medicinal Properties.—In proper doses it is a trustworthy diuretic ; but in large doses it is emetic and purgative. Scoparin in five-grain doses has also been given as a diuretic.

Official Preparations.

Decoctum Scoparii, from dried Broom Tops.
Succus Scoparii, from fresh Broom Tops.

---

2. INDIGOFERA TINCTORIA, *Linn.*

Indigo.

(Bentley and Trimen's ' Medicinal Plants,' vol. ii. plate 72.)

Habitat.—Its native country is probably Senegal and other parts of West Tropical Africa.   It is also apparently wild throughout India, where it is cultivated to an enormous extent, and also largely in Tropical Africa, Tropical America, &c.

Official Product and Name.—INDIGO :—a blue pigment prepared from various species of Indigofera, *Linn.*

Indigo.

Preparation and Commerce.—Indigo is prepared from the plants, which are cut down for that purpose just before the process of flowering, by a kind of fermentation.

It is chiefly imported from the East Indies ; but to some extent, also, from Guatemala and the northern parts of South America.

General Characters and Composition.—The description of these does not come within our province, but reference should be made to works on Chemistry.

Official Preparation and Uses.—Indigo has been introduced into the British Pharmacopœia solely for the preparation of Solution of Sulphate of Indigo, which is employed as a test for free chlorine in Liquor Sodæ Chlorinatæ and Acidum Hydrochloricum, as the colour of the solution is destroyed by free chlorine.

3. ASTRAGALUS GUMMIFER, *Labill.*

Tragacanth.

(Bentley and Trimen's ' Medicinal Plants,' vol. ii. plate 73.)

Habitat.—This shrub has a wide range, occurring on the Lebanon and Mount Hermon in Syria, and on the mountains of several parts of Armenia, Asia Minor, Northern Kurdistan, and Persia.

Official Product and Name.—TRAGACANTHA :—a gummy exudation obtained by making incisions in the stem of

Astragalus gummifer, *Labill.*; and some other species of Astragalus, *Linn.*

Tragacantha.

Tragacanth.

Nature, Collection, and Commerce.—Tragacanth is, what has been termed, a *degradation product*, owing, as it does, its production to the more or less complete transformation of the parenchymatous cells of the pith and medullary rays of the stem into a mucilaginous mass.

Tragacanth is produced in various parts of Asia Minor, Syria, Armenia, Kurdistan, and Persia. It exudes to some extent naturally, but the finest Tragacanth—*the flaky kind*—which is *alone official*, is obtained by making longitudinal incisions in the stem, when the juice exudes the whole length of each incision and dries in flakes. This kind of traga-canth is usually obtained from Smyrna, either direct, or occasionally by way of Constantinople ; hence its common name of *Smyrna Tragacanth.* The so-called *Syrian traga-canth* is improperly so termed, as it comes to us by way of Bagdad and the Persian Gulf, and is the produce of Kurdistan and Persia. It is flaky, like *Smyrna tragacanth*, but more translucent, and without any yellow tinge.

General Characters and Tests.—Flaky tragacanth is in white or somewhat yellowish flakes or leaf-like pieces, from one to three or more inches in length, and from a quarter to one inch broad. The pieces are thin, irregularly oblong or roundish, more or less curved, and marked on the surface by arched or concentric ridges ; they are more or less trans-lucent, tough, but rendered more pulverisable by a heat of 120° F. ; inodorous, and almost tasteless. Tragacanth is very sparingly soluble in cold water, but swelling into a gelatinous mass, which is tinged violet or blue by tincture of iodine. After maceration in cold water the fluid portion is not precipitated by the addition of rectified spirit.

Varieties of Tragacanth.—Besides the above *flaky traga-*

canth, other varieties are distinguished in commerce, as *vermiform* or *vermicelli tragacanth*, and *common tragacanth* or *tragacanth in sorts.*

Vermiform tragacanth is in narrow bands or string-like pieces, which are more or less curled or coiled. This kind is either the naturally exuded gum, or that which flows from artificial punctures.

Tragacanth in Sorts is essentially the gum which has exuded spontaneously. The pieces vary in size from that of a pea upwards ; they are brownish or yellowish in colour, rounded or botryoidal in form, and have a dull somewhat waxy appearance.

These inferior kinds of tragacanth are often adulterated with other gums, which are, in some cases, whitened previously by *white lead.*

Principal Constituents.—Tragacanth is essentially composed of two gummy principles, one soluble in water and resembling the arabin of gum acacia, but not identical with it, as its solution yields an abundant precipitate with the neutral acetate of lead ; and the other swelling in water, but not dissolving. The former is known as *soluble gum* or *arabin of tragacanth* ; the latter as *traganthin, adraganthin* or *bassorin.*

Medicinal Properties.—Emollient and demulcent, but employed medicinally, not so much for its own properties, as for its forming a good vehicle for more active and heavy medicines.

Official Preparations.

Confectio Sulphuris.
Glycerinum Tragacanthæ.
Mucilago Tragacanthæ.
Pulvis Opii Compositus, which is also used for the preparation of Confectio Opii.
Pulvis Tragacanthæ Compositus.

4. GLYCYRRHIZA GLABRA, *Linn.*

Liquorice.

(Bentley and Trimen's ' Medicinal Plants,' vol. ii. plate 74.)

Habitat.—A native of the warmer parts of the Mediterranean region, North Africa, Spain, Italy, Dalmatia, Greece and Syria, and extending also to the Danubian provinces, and South Russia, and to Asia Minor, Persia, and Afghanistan.

Official Part and Name.—GLYCYRRHIZÆ RADIX :—the root and subterranean stems or stolons, fresh and dried.

Glycyrrhizæ Radix.

Liquorice Root.

Cultivation, Preparation, and Commerce.—The plant is cultivated to some extent in England, and also in Spain, Russia, Germany, and certain other parts of Europe. The varieties more commonly known in this country are *English Liquorice Root*, *Spanish*, and *Russian* ; and some is also imported from Germany. The first variety may be obtained both in a fresh and dried state, the latter kinds only in a dried condition. In this country *liquorice root* is prepared for use as follows :—the long vertical descending true roots, together with the horizontal stolons, runners, or underground stems, are carefully dug up, washed, trimmed, and sorted ; and then either sold fresh in their entire state, or cut into pieces, which are usually of from three to four inches in length, and dried. In the latter case the thin brownish outer coat of the bark is sometimes first removed by scraping, when it is termed *peeled* or *decorticated.*

General Characters.—When *fresh* it is in long cylindrical pieces of varying thickness, smooth and yellowish-brown or somewhat reddish externally, yellow and juicy internally, very flexible, easily cut, and consisting of a thick cortical portion surrounding a central woody axis, which in the case of the stolons contains a small pith. It has a peculiar earthy and somewhat sickly odour ; and a strong peculiar sweet taste. When *dried* the pieces are either *peeled* or *unpeeled*. In the latter case the characters are essentially the same as those of the fresh root, except that it is somewhat darker, furrowed or wrinkled longitudinally, dry internally, and having a feebly acrid, and, in some cases, a slightly bitter taste combined with the characteristic sweetness; but when peeled it has a yellow colour externally, and there is no acridity.

Principal Constituents.—Liquorice root contains *sugar*, *asparagin*, *starch*, &c., and about 6 per cent. of a peculiar sweet unfermentable amorphous substance termed *glycyrrhisin*, which is a glucoside, resolvable into *glucose* and a bitter amorphous substance called *glycyrretin*.

Medicinal Properties.—Liquorice root possesses demulcent properties; but it is principally used as a flavouring adjunct to nauseous medicines, &c., and for other pharmaceutical purposes.

Official Preparations.

Confectio Terebinthinæ.	Infusum Lini.
Decoctum Sarsæ Compositum.	Pilula Hydrargyri.
	Pilula Ferri Iodidi.
Extractum Glycyrrhizæ.	Pulvis Glycyrrhizæ Compositus.
Extractum Glycyrrhizæ Liquidum.	

Extractum Glycyrrhizæ and Extractum Glycyrrhizæ Liquidum are also contained in several other official preparations.

5. MUCUNA PRURIENS, *DC.*
Cowhage.

Synonym.—DOLICHOS PRURIENS, *Linn.*

(Bentley and Trimen's ' Medicinal Plants,' vol. ii. plate 78.)

Habitat.—It is extensively distributed over the Peninsula of India, where it appears to be a native. It also occurs in a semi-wild or cultivated state throughout the tropical regions of both hemispheres.

Part Used and Name.—MUCUNA:—the hairs covering the legumes.

(*Not Official.*)

Mucuna.

Cowhage.

General Characters.—The legumes with the hairs attached are imported from the West Indies. These legumes are from two to about four inches in length, and about half an inch in breadth, somewhat compressed, falcately curved at each end, and contain four or five seeds. Externally they have a dark rich-brown colour, and are densely covered with stiff brownish-red hairs which point backwards, and each about one-tenth of an inch in length. These hairs are readily detached, and constitute the substance termed *cowhage* or *cow-itch*. When incautiously touched they penetrate the skin, and cause intolerable itching. Each hair when examined with a magnifying lens is seen to consist of a single cell which is conical in form, and acute and somewhat barbed at the apex.

Principal Constituents.—Most of the hairs are filled with air, others contain a granular matter, of which *tannic acid* appears to be the principal constituent.

Medicinal Properties.—Cowhage is a mechanical anthelmintic when administered internally; and a local stimulant when applied externally.

---

6. PHYSOSTIGMA VENENOSUM, *Balf.*

Calabar Bean.

(Bentley and Trimen's ' Medicinal Plants,' vol. ii. plate 8c.)

Habitat.—A native of West Tropical Africa, near the mouths of the Niger and the Old Calabar River in the Gulf of Guinea.

Official Part and Name.—PHYSOSTIGMATIS SEMEN :— the dried seed.

Physostigmatis Semen.

Calabar Bean.

Synonym.—Physostigmatis Faba.

Commerce.—These seeds are imported from West Tropical Africa.

General Characters and Test.—Calabar beans (*fig.* 16, *a*) are from about one inch to one inch and a quarter long, about three-quarters of an inch broad, half an inch or more in thickness, and average about sixty-six grains in weight. In form they are oblong and more or less reniform, being straight or somewhat concave on one side, and convex on the other. The convex side is marked by a long broad blackish furrow (*hilum*) with raised edges, which extends along its whole length; this furrow is also traversed by a central ridge (*raphe*), and presents a small depression or aperture (*micropyle*) near one end. The testa is hard, roughish, brittle, and usually of a deep chocolate-brown colour, except on the raised edges of the furrow, where it is lighter; but sometimes the colour is brownish-red. The testa adheres closely

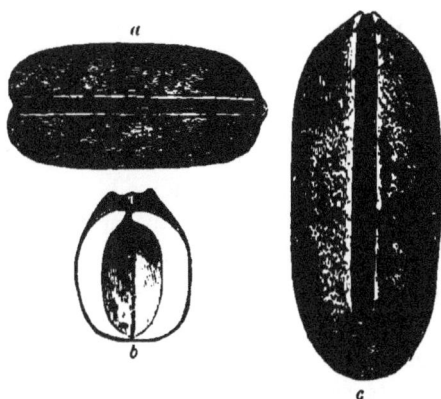

Fig. 16.—*a*. Physostigmatis Semen, *Calabar Bean.* *b*. Transverse section of a seed. *c*. Seed of *Physostigma cylindrosum.* (After Holmes.)

to the nucleus, and principally consists of two hard white brittle cotyledons, separated from each other by a somewhat large cavity (*fig.* 16, *b*). The seed has no odour, and no marked taste beyond that of an ordinary bean. It yields its virtues to alcohol, and imperfectly to water. The coty-

I

ledons when moistened with solution of potash acquire a permanent pale-yellow colour.

The above description especially applies to the seeds of *Physostigma venenosum*, but in commerce, as noticed by Holmes in 1879, the seeds of another species, or probably only a variety, of *Physostigma*, termed *Physostigma cylindrosum*, were found. The latter seeds are somewhat longer (*fig.* 16, *c*), more cylindrical in form, and the hilum or furrow does not extend more than about three-fourths along the convex side of the seed. It is probable that there is no essential difference in the properties of the two seeds, but as we have no positive evidence on this point, the description in the pharmacopœia is especially drawn to exclude the cylindrical seeds.

Adulterations.—The seeds of one or more species of *Mucuna*, and of the Oil Palm (*Elais guineensis*, Jacq.), have been found mixed with Calabar beans; but the characters of these are so well marked that no difficulty can arise in detecting them and any other seeds under such circumstances.

Principal Constituents.—The properties of Calabar beans are essentially due to a powerful alkaloid called *physostigmine* or *eserine*. It is principally contained in the nucleus, but also to a small extent in the testa. It is readily soluble in alcohol, but only slightly so in water. This alkaloid *is official.* Another alkaloid, *Calabarine*, which produces tetanic effects like strychnine, has also been found in these seeds.

Medicinal Properties.—Calabar bean is a powerful sedative of the spinal cord, and in over-doses poisonous. It is but rarely administered internally; but when applied locally to the eye it contracts the pupils, and hence it is much employed in certain diseases and injuries of that organ, and in operations. The alkaloid physostigmine is used for like purposes. The dose of Calabar bean, in powder, is from one to four grains.

Official Preparations.

Extractum Physostigmatis. *Dose.*—$\frac{1}{16}$ to $\frac{1}{4}$ grain.

From this extract the official alkaloid Physostigmina is directed to be obtained ; and from this the official Lamellæ Physostigminæ are prepared. Each disc of the latter weighing about $\frac{1}{50}$ grain, and containing about $\frac{1}{1000}$ grain of physostigmine.

7. PTEROCARPUS MARSUPIUM, *Roxb.*
Malabar Kino.

(Bentley and Trimen's ' Medicinal Plants,' vol. ii. plate 81.)

Habitat.—Forests of Central and Southern India. It is also found in Ceylon.

Official Product and Name.—KINO:—the juice obtained from incisions made in the trunk, inspissated without artificial heat.

Kino.
Kino.

Varieties and Commerce.—Besides the official kino, other kinds are known to pharmacologists : as *Butea* or *Bengal Kino*, from *Butea frondosa*, Roxb., and other allied species of *Butea* ; *African* or *Gambia Kino*, from *Pterocarpus erinaceus*, Poiret; *Botany Bay*, *Eucalyptus*, or *Australian Kino*, from numerous species of *Eucalyptus* ; and others. Hence the official kind is commonly distinguished under the name of *Malabar* or *East India Kino*, from its geographical source being the forests of the Malabar Coast of India.

Extraction.—Kino is obtained as follows :—A perpendicular incision is first made in the trunk, and then lateral incisions leading into it, when the juice exudes, and is collected by placing a suitable vessel at the lower end of the main incision. This juice is dried by exposure simply to sun and air, and then broken up into small fragments.

General Characters.—Commercial kino is found in

small angular very brittle glistening reddish-black fragments, which are opaque when entire, but when in thin laminæ and at their edges, they are transparent and of a ruby-red colour. Kino is inodorous, but has a very astringent taste, and when chewed it sticks to the teeth and tinges the saliva blood-red. It is almost entirely soluble in rectified spirit, but yields little or nothing to ether, and is only partially dissolved by cold water.

Principal Constituents.—Its essential constituents are *pyrocatechin, kino-tannic acid*, and *kino-red.*

Medicinal Properties.—Powerfully astringent ; and may be used, therefore, both internally and externally, in all cases where tannic acid is indicated.

Official Preparations.

Pulvis Catechu Compositus. | Pulvis Kino Compositus.
Tinctura Kino.

8. PTEROCARPUS SANTALINUS, *Linn. fil.*

Red Sanders Wood.

(Bentley and Trimen's ' Medicinal Plants,' vol. ii. plate 82.)

Habitat.—It is a native of the southern part of the Indian Peninsula ; and is also said to occur in Mindanao in the Southern Philippines.

Official Part and Name.—PTEROCARPI LIGNUM :—the sliced or rasped heart-wood.

Pterocarpi Lignum.

Red Sandal Wood.

Synonym.—Red Sanders Wood.

Commerce.—It is chiefly imported from Madras in the form of logs or billets, which are principally obtained from the lower part of the trunk and from the thickest roots, and from which the bark and sap-wood have been removed.

General Characters.—The logs are of various sizes and lengths, but usually from about three to five feet long, and although sometimes as thick as a man's thigh, they are commonly much smaller. They are roundish, irregular or somewhat angular, heavy, dense, dark reddish-brown or blackish-brown externally, and internally, if cut transversely, deep blood-red variegated with zones of a lighter red colour. Red Sandal Wood is usually found in the pharmacies in the form of raspings or small chips, which have a deep reddish-brown or purplish-red colour, a very slightly astringent taste, and are almost inodorous except when rubbed, when they have a faint peculiar smell.

Principal Constituents.—The chief constituent is the colouring principle, termed *santalic acid* or *santalin*, which is insoluble in water either hot or cold, but readily soluble in alcohol, ether, and alkaline solutions.

Medicinal Properties.—It possesses very slight astringent properties; but in this country it is only used in pharmacy for colouring the Compound Tincture of Lavender.

Official Preparation.—Tinctura Lavandulæ Composita, which is also used in the preparation of Liquor Arsenicalis.

.

9. **ANDIRA ARAROBA**, *Aguiar.*

Araroba.

('Pharmaceutical Journal,' 3rd ser. vol. x. page 43.)

Habitat.—Native of Bahia and Brazil.

Official Product and Name. — CHRYSAROBINUM: — the medullary matter of the stem and branches; dried, powdered, and purified; containing more or less chrysophanic acid according to age and condition, and yielding much chrysophanic acid by oxidation.

Chrysarobinum.

Chrysarobin.

Synonyms.—Araroba Powder. Goa Powder.

Commerce.—It is imported from Brazil.

General Characters and Tests.—Commercial chrysarobin, as purified by solvents, occurs as a light brownish-yellow, minutely crystalline powder, tasteless and inodorous. It is very sparingly soluble in water, but almost entirely soluble in 150 parts of hot rectified spirit. On heating it melts and partially sublimes in yellow vapours, leaving a charred residue, which entirely disappears on ignition in air. It dissolves in sulphuric acid to form a yellow to orange-red solution, and in solution of caustic potash to form a yellow to reddish fluorescent solution which becomes carmine by absorption of oxygen from the air.

Principal Constituents.—Its essential constituent is *chrysophanic acid.*

Medicinal Properties.—When used externally it is a powerful stimulant and parasiticide in many skin affections. It has also been administered internally in doses of from one-sixth to half a grain, but it is liable to produce severe purgation.

Official Preparation.—Unguentum Chrysarobini.

---

10. MYROXYLON PEREIRÆ, *Klotzsch.*

Balsam of Peru.

Synonym.—TOLUIFERA BALSAMUM, VAR., *Baill.*

(Bentley and Trimen's ' Medicinal Plants,' vol. ii. plate 83.)

Habitat.—It is found on the Sonsonate coast, or ' Balsam Coast,' of the State of San Salvador, Central America, formerly part of Guatemala.

Official Product and Name.—BALSAMUM PERUVIANUM:—
a balsam exuded from the trunk, after the bark has been
beaten, scorched, and removed.

Balsamum Peruvianum.

Balsam of Peru.

Extraction and Commerce.—Balsam of Peru is exclusively
the produce of the State of Salvador in Central America.
It is obtained as follows :—The trees are beaten on the four
sides with some blunt instrument, as the back of an axe,
until the bark is loosened, the four intermediate strips being
left untouched in order not to destroy the life of the tree,
and to be similarly treated the following year. The bruised
bark soon cracks longitudinally, is readily separable from
the wood, and is sticky on its under surface, as well as the
wood itself, from the exudation of a small quantity of
fragrant balsam, but this is in too small amount to be worth
collecting. Hence, in order to promote a more abundant
flow, the injured bark, five or six days after it has been
beaten, is charred by applying to it lighted torches or pieces
of burning wood. In a few days more the charred bark
falls off, or is removed, and the trunk commences to exude
the balsam more freely. Pieces of rag are then placed on
the bare portions of the trunk for the purpose of collecting
the balsam ; and as these become saturated with it, they are
collected and thrown into an earthenware boiler nearly filled
with water, and then, by gently boiling, the rags become
quite clean, and the balsam, which at first is of a light
yellowish colour, becomes dark coloured and heavy, and
sinks to the bottom ; and when the water has cooled, it is
poured off, and the balsam put into gourds of different sizes
and sent to the market.

Balsam of Peru is now chiefly exported in tin canisters
or drums, and principally by way of Acajutla on the Pacific
coast ; but also to some extent from Belize and other ports
on the Atlantic side of Central America.

General Characters and Tests.—Balsam of Peru is a liquid somewhat less viscid than treacle, and when seen in bulk, nearly black in colour, but when examined in thin layers, it is transparent, and deep orange-brown or reddish-brown. Its odour is agreeably balsamic, more especially when thinly spread out on paper and heated. It has a warm and bitterish taste, and when swallowed it leaves a disagreeable burning sensation in the throat. It is soluble in chloroform or rectified spirit, but insoluble in water. Its specific gravity varies from 1·137 to 1·150. Ten drops triturated with six grains of slaked lime produces a permanently soft mixture; and the mixture, on being warmed until all volatile matter is given off and until charring commences, gives no fatty odour. It should not diminish in volume when shaken with an equal bulk of water.

Principal Constituents.—It is essentially composed of about 38 per cent. of an odourless and tasteless *resin*, which, by destructive distillation, furnishes *benzoic acid, styrol,* and *toluol*; and of about 60 per cent. of *cinnamein* or *benzylic cinnamate*, resolvable into *cinnamic acid* and *benzylic alcohol.*

Medicinal Properties.—Stimulant and expectorant, when given internally, but now rarely used. Locally applied it is stimulant and parasiticide.

11. MYROXYLON TOLUIFERA, *H. B. and K.*
Balsam of Tolu.

Synonym.—Toluifera Balsamum, *Mill.*

(Bentley and Trimen's 'Medicinal Plants,' vol. ii. plate 84.)

Habitat.—A native of Venezuela and New Granada; and probably also of Peru, Ecuador, and Brazil.

Official Product and Name.—Balsamum Tolutanum :— a balsam which exudes from the trunk after incisions have been made in the bark.

Balsamum Tolutanum.

Balsam of Tolu.

Extraction and Commerce.—Balsam of Tolu is the produce of New Granada, and is obtained as follows :—Two
sloping incisions are made quite through the bark to the
wood, so as to meet at an acute angle at their lower ends ;
the bark and wood are then slightly hollowed out immediately
below this V-shaped incision, and a small calabash, about
the size and shape of a deep tea-cup, is inserted in order to
receive the balsam as it exudes from the incision. Numerous calabashes are inserted in a similar manner over the
trunk at close intervals, and beneath similar incisions ; and
from time to time, as may be necessary, the contents of
these cups are poured into flask-shaped bags of raw hide,
and the calabashes replaced in their former positions to fill
again. The balsam is then sent down to the ports on the
river Magdalena, where it is transferred to cylindrical tins,
which commonly contain about ten pounds, or sometimes
more, and exported to Europe and other parts.

General Characters.—When first imported it is a soft
and tenacious solid, which varies, however, with the temperature ; but by keeping, it becomes gradually harder, and
then, in cold weather, is brittle like resin; and when old
it exhibits when broken a crystalline appearance. In thin
films it is transparent and of a yellowish-brown colour ; and
when pressed between two pieces of glass with the aid of
heat, and then examined with a lens, it exhibits an abund-
ance of crystals of cinnamic acid. It has a highly fragrant
odour, especially when warmed; and a somewhat aromatic,
slightly acid, not unpleasant taste. It is soluble in rectified
spirit, and the solution has an acid reaction.

Principal Constituents.—Its constituents are *amorphous
resin, cinnamic* and *benzoic acids*, about 1 per cent. of a
volatile oil obtained by distilling the balsam with water
(*tolene*), and *benzylic ethers* of both cinnamic and benzoic

acids. By destructive distillation it affords the same substances as balsam of peru under similar conditions.

Medicinal Properties.—It has similar stimulant and expectorant properties to balsam of peru, but it is much more
frequently employed medicinally. It is also frequently used
as an agreeable flavouring adjunct to pectoral mixtures.

Official Preparations.

Pilula Phosphori.   Tinctura Benzoinæ Composita.
Syrupus Tolutanus. ɪ Tinctura Tolutana.

Tinctura Tolutana is also an ingredient in—

Trochisci Acidi Tannici.   Trochisci Morphinæ et
Trochisci Morphinæ.   Ipecacuanhæ.
                    Trochisci Opii.

---

12. HÆMATOXYLON CAMPECHIANUM, *Linn.*
Logwood.

(Bentley and Trimen's ' Medicinal Plants,' vol. ii. plate 86)

Habitat.—It is a native of Tropical America, especially
the shores of the Bay of Campeachy, Honduras, and
Columbia.   It is also now naturalised in Jamaica and other
of the West Indian Islands.

Official Part and Name.—HÆMATOXYLI LIGNUM :—the
sliced heart-wood.

Hæmatoxyli Lignum.
Logwood.

Production and Commerce.—When about ten years old,
the trees are felled ; after which, the bark and sap-wood are
removed, and the heart-wood cut into logs about three feet
ong, in which form it is exported. It is obtained from
Central America and the West Indies, and is distinguished

in the London market according to its geographical source, as Campeachy, Honduras, St. Domingo, and Jamaica Logwood ; the former being the most esteemed.

General Characters.—The logs (*heart-wood*) are hard and heavy, having a specific gravity of about 1·057. Externally, by exposure to air and moisture, they become dark purplish or blackish-red in colour; internally they are brownish-red. The chips, in which form logwood is commonly found in the pharmacies, and officially directed to be used, have a dark reddish-brown colour, and often a greenish lustre. Their odour is faint and agreeable, and has been compared to violets and sea-weed ; and when chewed, they have a sweetish astringent taste, and impart to the saliva a brilliant dark reddish-pink colour.

Principal Constituents.—Its chief constituent is *hæmatoxylin*, which, when pure, is in colourless crystals ; these are soluble in hot water or alcohol, and have a sweet taste. Under the influence of alkalies. when exposed to the air, hæmatoxylin becomes red. The green hue which is frequently observed on logwood chips is produced by the transformation of hæmatoxylin, under the influence of ammonia and oxygen, into a crystalline substance termed *hæmatein.* *Tannic acid* is also a constituent of logwood.

Medicinal Properties.—Logwood is a useful astringent.

Official Preparations.

Decoctum Hæmatoxyli. | Extractum Hæmatoxyli.

---

13. CASSIA FISTULA, *Linn.*

Purging Cassia.

(Bentley and Trimen's ' Medicinal Plants,' vol. ii. plate 87.)

Habitat.—A true native of India and other parts of Tropical Asia. It is also now grown, or possibly indigenous, in Egypt and other parts of Africa; and cultivated in the West Indies, especially Jamaica, and in Brazil.

Official Part and Name.— CASSIÆ PULPA :—the pulp obtained from the recently imported pods.

General Characters of Cassia Fistula.—The pods or fruits, which are known under the name of '*Cassia Fistula*,' are botanically characterised as lomentaceous legumes. They vary in length from one and a half to two feet, and from three-quarters to one inch in diameter; are shortly stalked, pointed at the apex, blackish-brown in colour, marked with faint transverse depressions and veins, very hard and indehiscent, but the sutures distinctly marked by two nearly smooth longitudinal bands running along their whole length. They are divided internally by thin flat transverse partitions or dissepiments into numerous cells (*fig.* 17), each of which contains a solitary flattish-ovoid hard smooth and shining reddish-brown seed; this in the recent fruit is entirely imbedded in a soft viscid pulp, but in the pods of commerce the pulp is only found lining the inner surface of the cells, or even simply on the partitions, the seeds being free from it. As the pulp is alone official, those pods should be selected which are

FIG. 17. Vertical section of a portion of the pod or fruit (*lomentaceous legume*) of *Cassia Fistula*, showing the transverse dissepiments by which it is divided into a corresponding number of cells, in each of which is a solitary seed, surrounded by pulp.

heavy and do not rattle when shaken. The pods should yield about 30 per cent. of pulp.

Cassiæ Pulpa.

Cassia Pulp.

Commerce.—The pods are principally imported from the West Indies, but likewise to some extent from the East Indies. The pulp is also sometimes derived *per se* from the East Indies, but as it becomes acid by exposure to the air, it is officially directed to be obtained from the recently imported pods, and it should be used as soon as possible after its extraction.

General Characters.—The pulp has a viscid consistence, a blackish-brown colour, shining appearance, sweet taste, and somewhat sickly odour, but somewhat resembling prunes. When obtained separately the pulp commonly contains more or less of the seeds and partitions or dissepiments; these should be removed when it is used for pharmaceutical and medicinal purposes.

Principal Constituents.—There is no peculiar principle in the pulp, but it contains about 60 per cent. of *sugar*, together with some *mucilage, pectin, albuminoids*, and other unimportant constituents.

Medicinal Properties.—Laxative, and in large doses purgative; but in the latter case, when used alone, it commonly causes nausea, griping, and flatulence.

Official Preparation.—Confectio Sennæ.

CASSIA GRANDIS.

(*Not Official.*)

The pods of this plant, which is a native of Brazil and Central America, are sometimes used in veterinary practice, and are, therefore, termed Horse Cassia. They are commonly longer than Cassia Fistula, and of greater diameter, being frequently as much as one and a half inch broad. They are also readily distinguished by being laterally compressed, very rough externally, and by being marked by three longitudinal prominent ridges instead of two nearly smooth bands as in Cassia Fistula.

14. CASSIA ACUTIFOLIA, *Delile.*

Alexandrian Senna. Nubian Senna.

Synonym.—CASSIA LANCEOLATA, *Nectoux.*

(Bentley and Trimen's 'Medicinal Plants,' vol. ii. plate 90.)

Habitat.—A native of Tropical Africa, principally Nubia and especially the southern districts Sennaar and Kordofan.

Official Part and Name.—SENNA ALEXANDRINA :—the dried leaflets.

Senna Alexandrina.

Alexandrian Senna.

Collection and Commerce.—This Senna, which is termed *Senna Jebeli* or *Mountain Senna*, is essentially collected and dried for use in Nubia. It is then commonly forwarded by way of Assouan and Darao, and thence by the Nile to Cairo, from whence it reaches Alexandria, from which port it is ultimately exported and derives its name.

General Characters. — The leaflets are ordinarily about one inch long, though varying in this respect from about three-quarters of an inch to nearly one inch and a quarter (*fig.* 18). They are lanceolate or oval-lanceolate in shape, acute-pointed, unequal at the base, commonly entire, brittle, pale yellowish-green, opaque, evidently veined on the lower surface, and slightly pubescent, more especially on the midrib, or nearly smooth. Their odour is faint and peculiar,

FIG. 18.—Leaflet of *Cassia acutifolia.*

although somewhat tea-like ; and their taste is mucilaginous, nauseous, and sickly, more especially in that of a watery infusion.

Adulterations.—It has been frequently imported in a more or less contaminated condition, in which case, as directed in the pharmacopœia, the true senna leaflets should be carefully separated from all extraneous matters. Thus we may find (although Alexandrian Senna is now commonly of much better quality than formerly) a variable proportion of the leaf-stalks, flowers, broken twigs, and legumes of the same plant ; also, those of *Cassia obovata*, Colladon ; the leaves, flowers, and fruits of *Solenostemma Argel*, Hayne ; and

FIG. 19. — Leaflet of *Cassia obovata*.   FIG. 20. — Leaflet of *Tephrosia Apollinea*.   FIG. 21.—Leaf of *Soleno-stemma Argel*.

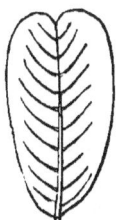

very rarely the leaflets and legumes of *Tephrosia Apollinea*, Delile ; together with date stones and various other extraneous matters. Many of these contaminations may be readily separated from the true senna leaflets by sifting, fanning, and picking ; but if the official senna be mixed with other senna leaflets, or with the leaves of other plants, these are frequently left behind, and hence it is neces-sary for us to describe the principal characters of these adulterations. Thus, in the first place, the leaflets of *Cassia obovata*—which were official in the Pharmacopœia of 1867 as one of the constituents of Alexandrian Senna—are some-times found in varying proportions. They are readily distinguished from the leaflets of *Cassia acutifolia* by their oblong-obovate outline (*fig.* 19) and somewhat mucronate apex. These obovate leaflets are now regarded as very inferior to the official senna leaflets; they form the *Wild Senna* or *Senna Baladi* of the Arabs. A more serious admixture, however, is that of *Argel leaves*, from *Solenostemma Argel.* These may be known from all senna leaflets by their paler colour, more leathery texture, less conspicuous veins (*fig.* 21), by being equal-sided at the base, and by their very bitter taste. The *Tephrosia leaflets* (*fig.* 20) are very rarely if ever found at the present day, but are easily distinguished by being usually folded on their midrib, their obovate-oblong outline, emarginate apex, silvery or silky appearance, and by being equal-sided at the base.

Principal Constituents.—*Cathartic acid*, a glucoside acid, which is resolved by mineral acids into *glucose* and *cathartogenic acid*, is regarded as the essential active principle. When pure it is insoluble in water or alcohol, but soluble in ether; it exists, however, in senna, combined with calcium and magnesium, and in this form it is readily soluble in water, although still insoluble in alcohol. Senna, also, contains a saccharine crystalline substance, termed *catharto-mannite* ; a yellow colouring principle, which is said to be identical with *chrysophanic acid*, although sometimes termed *chrysoretin* ; *mucilage*, &c. Two bitter principles, *sennacrol* and *sennapicrin*, are also said to be constituents of senna.

Medicinal Properties. — Purgative, for which purpose senna is extensively employed.

Official Preparations.

Confectio Sennæ.	Pulvis Glycyrrhizæ Compositus.
Infusum Sennæ.	Syrupus Sennæ.
Tinctura Sennæ.	

Infusum Sennæ and Tinctura Sennæ are also used in the preparation of Mistura Sennæ Composita.

15. CASSIA ANGUSTIFOLIA, *Vahl.*

Arabian Senna. Tinnevelly Senna.

Synonym.—CASSIA ELONGATA, *Lem-Lisanc.*

(Bentley and Trimen's ' Medicinal Plants,' vol. ii. plate 91.)

Habitat.—It is a common wild plant throughout the Yemen and Hadramaut provinces of Southern Arabia. It is also found on the Somali coast in Eastern Tropical Africa, in Scinde and the Punjaub. It is cultivated in Southern India.

Official Part and Name.—SENNA INDICA :—the dried leaflets. From plants cultivated in Southern India.

Varieties.—There are two varieties of Indian Senna, namely, *Arabian, Mocha,* or *Bombay Senna;* and *East Indian* or *Tinnevelly Senna,* the latter being *alone official.—Arabian or Bombay Senna.*—This kind of senna is collected in Southern Arabia, from whence it is exported by way of the Red Sea ports to Bombay, and thence to Europe and other parts. As imported, the leaflets are often broken and mixed with the stalks, flowers, and legumes of the same plant. The leaflets are narrowly lanceolate or obovate-lanceolate, unequal at the base, tapering towards the apex, from one to two inches long, smooth or very slightly pubescent, and frequently brown and decayed from imperfect drying. They are readily distinguished from the leaflets of Alexandrian senna by their length and narrowness, so that in France, from their pike-like shape, they form what is termed *séné de la pique.* It is regarded as a very inferior senna.

Senna Indica.

East Indian Senna.

Synonym.—Tinnevelly Senna.

Cultivation and Commerce.—This kind of senna is, like the Arabian senna, derived from *Cassia angustifolia,* but

K

from cultivation in Southern India the plant grows more luxuriantly than in its wild state in the drier regions of Arabia, and forms the official East Indian or Tinnevelly Senna, which is commonly exported from Tuticorin in the extreme South of India. As imported it is free from all other leaves and extraneous matters of any kind.

General Characters.—The leaflets of this kind of senna are from about one to two inches long, and sometimes half an inch broad at their widest part (*fig.* 22). They are lanceolate in outline, acute-pointed, un-equal-sided at the base, thin, unbroken ; yellowish-green and smooth above, but some-what of a duller tint on their under surface, and glabrous or slightly pubescent from short adpressed hairs. In odour and taste they closely resemble Alexandrian senna. Tinne-velly senna, from its careful preparation, freedom from adulteration, &c., is regarded as a fine variety, although its quality, so far as regards its size, has fallen off of late years.

FIG. 22.—Leaflet of a cultivated plant of *Cassia angustifolia.*

Principal Constituents and Medicinal Properties.—These are given under Alexandrian Senna.

Official Preparations.

This Senna may be used in place of Alexandrian Senna.

16. TAMARINDUS INDICA, *Linn.*
Tamarind.

(Bentley and Trimen's 'Medicinal Plants,' vol. ii. plate 92.)

Habitat.—It is truly indigenous in Tropical Africa. It also grows throughout India, in the Philippines, Java, etc. :

and has been introduced into the West Indian Islands, Brazil, Central America, etc.

Official Part and Name.—TAMARINDUS :—the preserved pulp of the fruit.

General Characters of the Fruit.—The fruit is an indehiscent legume, more or less oblong in form, about one inch broad, and from three to six inches long, slightly compressed, somewhat curved, smooth, and supported on a woody stalk. The *epicarp* is thin, brittle, and of a pale chocolate-brown colour. Within the epicarp is a firm juicy brownish or reddish-brown pulp or *sarcocarp*, on the surface of which are three tough branching woody veins, which extend from the stalk to the apex of the fruit ; and, imbedded in the pulp, there are a variable number of seeds, usually from three to ten, each enclosed in a tough membranous cell or *endocarp*.

Tamarindus.

Tamarind.

Kinds, Preparation, and Commerce.—There are two kinds of Tamarinds found in commerce, which are known from their sources, respectively, as *West Indian Tamarinds* and *East Indian Tamarinds*. The former are preserved as follows : The shell or epicarp being removed from the ripe fruits, alternate layers of the shelled fruits and powdered sugar are placed in a cask or jar, and boiling water poured over them till the cask or jar is full ; or the shelled fruits are simply placed in layers and boiling syrup poured over them. ˙ In the East Indies, although tamarinds are occasionally preserved as above with sugar, the usual practice is simply to remove the shell from the fruits, and press the remaining portions together into a mass. *The pulp preserved with sugar* as commonly exported from the West Indies, *is alone official in this country*, and is the form in which it is usually found in the pharmacies.

1. *West Indian Tamarinds.*—These, which are also

K 2

known as *Brown* or *Red Tamarinds*, are characterised in the pharmacopœia as follows :—' A reddish-brown moist sugary mass, enclosing strong branched fibres, and brown shining seeds, each enclosed in a tough membranous coat. Taste agreeable, refreshing, sub-acid.' As these tamarinds are sometimes said to be prepared in copper vessels, the following test of their purity from such contamination is given in the pharmacopœia :—' A piece of bright iron, left in contact with the pulp for an hour, does not exhibit any deposit of copper.'

2. *East Indian or Black Tamarinds.*—These, as commonly imported, are preserved without sugar, and are found in the market as a firm clammy dark-brown or black mass, which consists of the pulp and seeds, mixed with strong branched fibres, and some fragments of the shell. They have a very acid taste.

Principal Constituents.—There is no known principle present in tamarinds to account for their laxative action, but unsweetened tamarinds contain *tartaric, citric,* and *acetic acids,* either in a free state, or more generally combined with potash. *Grape sugar* and *pectin* are also constituents.

Medicinal Properties.—Slightly laxative and refrigerant.

Official Preparation.—Confectio Sennæ.

---

17. COPAIFERA LANGSDORFFII, *Desf.*
Copaiba.

(Bentley and Trimen's ' Medicinal Plants,' vol. ii. plate 93.)

Habitat.—This species is found growing wild over a vast area in Brazil.

Official Product and Name.—COPAIBA :—the oleo-resin obtained by cutting deeply or boring into the trunk of Copaifera Langsdorffii, *Desf.;* and other species of Copaifera, *Linn.*

Copaiba.

Copaiva or Copaiba.

Extraction and Commerce.—Copaiba is collected chiefly in Brazil in the valleys of the Amazon and its tributaries, from whence it is exported chiefly from Para. Copaiba is also exported in large quantities from Maracaibo ; this kind being obtained in Venezuela, and probably from *Copaifera officinalis*, Linn. Some is also collected on the banks of the Orinoco, and forwarded to Europe, etc., by way of Angustura and Trinidad. Copaiba is also exported, to some extent, from other places. It frequently reaches England by way of Havre and New York. The mode in which the oleo-resin is obtained is by boring deeply into the heart-wood of the trees, or more commonly by cutting a hole or chamber in their trunks near the base. The copaiba then usually flows readily from the wounded trunks, and in large quantities ; it is at first thin, clear, and colourless, but it soon becomes thicker and acquires a yellowish tinge. It is commonly exported in barrels, and the different kinds of copaiba are known in commerce as Para, Maranham, Maracaibo, West Indian, &c.

General Characters.—*Copaiva, Copaiba,* or *Capivi,* is a more or less viscid liquid, generally transparent and not fluorescent; but some kinds are opalescent and occasionally slightly fluorescent. Its colour varies from a pale yellow to a light golden-brown; its odour is peculiar, aromatic, and not unpleasant ; and it has a persistent acrid somewhat bitter taste. Its specific gravity varies from 0·940 to about 0·993, in all cases being lighter than water. Copaiba becomes denser and acquires a deeper colour by keeping, owing partly to the volatilisation, and partly to the oxidation, of its volatile oil. In its action on polarised light the different kinds of copaiba vary not only in degree, but also in direction, some being dextrogyre, others levogyre.

Tests.—The following tests of copaiba are given in the pharmacopœia :—A small quantity heated until all volatile oil is removed yields a residue which when cold is hard, and, generally, easily rubbed to powder ; and the oil volatilised during the operation does not smell of turpentine. It is almost entirely soluble in absolute alcohol, and in four times its bulk of petroleum spirit, the latter solution only yielding a filmy deposit on standing.

Varieties.—The several varieties known as Para, Maranham, Maracaibo, West Indian, &c., although distinguished by experienced judges from their differences of colour, density, odour, taste, degree of transparency, proportion of volatile oil, &c., have no sufficiently definite characters, so as to be accurately described by pharmacologists.

Adulterations.—Copaiba is not unfrequently adulterated with oil of turpentine, castor oil, and other fixed oils. *Wood Oil* or *Gurjun Balsam*, from *Dipterocarpus turbinatus*, and other species of *Dipterocarpus*, and other substances, are also sometimes mixed with, or substituted for copaiba. Oil of turpentine may be recognised by its odour on heating ; fixed oils, by their being mostly insoluble in alcohol, and in the case of castor oil, which is soluble in that menstruum, by gently heating the suspected copaiba with moistened magnesia, lime, or baryta, when, if free from adulteration, it forms a more or less stiff compound. Wood oil may be distinguished by gently heating the copaiba in a closed glass-tube, when, if pure, it becomes more fluid ; while wood oil, when heated to about 270° Fahr., gelatinises, and on cooling does not resume its former fluidity. Copaiba should also answer to the other tests of the pharmacopœia.

Principal Constituents.—The essential constituent is a *volatile oil*, which is colourless or pale yellow, and has the odour and taste of copaiba ; this oil *is official*. Its amount varies in different samples according to their age and botanical source, from about 40 to 60 per cent. After the oil has been removed by distillation, an *acid brown amor-*

phous resin remains ; hence copaiba is an oleo resin, and its common name *Balsam of Copaiba* is therefore incorrect, as it contains neither benzoic or cinnamic acid, the presence of one or the other of which is necessary to constitute a balsam. Although the ordinary acid resin of copaiba is amorphous, resin crystalline acids may be obtained from different kinds of copaiba, which are termed *copaivic acid*, *oxycopaivic acid*, and *metacopaivic acid.*

Medicinal Properties.—Copaiba is a stimulant like the ordinary turpentines, but its action is more especially on the mucous membranes, and particularly on those of the genitourinary organs and the lungs.   *Dose.*—$\frac{1}{2}$ to 1 fluid drachm.

The official volatile oil has similar properties to that of the oleo-resin. The acid resin has also diuretic properties.

Official Preparation.—Oleum Copaibæ. The oil distilled from copaiva.   *Dose.*—5 to 20 minims.

---

18. ACACIA SENEGAL, *Willd.*
Gum Acacia.

Synonym.—ACACIA VEREK, *Guill. & Perr.*

Bentley and Trimen's ' Medicinal Plants,' vol. ii. plate 94.)

Habitat.—This species is a native of Senegal, in Western Africa, and is also found in Southern Nubia, Kordofan, and in the region of the Atbara, in Eastern Africa.

Official Product and Name.—ACACIÆ GUMMI:—a gummy exudation from the stem and branches of A. Senegal, *Willd.*; and from other species of Acacia, *Willd.*

Acaciæ Gummi.
Gum Acacia.

Collection, Commerce, and Varieties.—Gum Acacia commonly exudes spontaneously, but in some districts the outflow is facilitated by incisions. There are several varieties known in the London market, the more important being *Kordofan, Picked Turkey* or *White Sennaar Gum ; Senegal Gum; Suakim Gum; Morocco, Mogador,* or *Brown Barbary Gum; Cape Gum; East Indian Gum ;* and *Australian Gum.* Kordofan Gum, which is known in Egypt as *Hashabi Gum,* is the finest kind, and is, as its name implies, produced in Kordofan, and its botanical source is *Acacia Senegal.* Senegal Gum is also the produce of *A. Senegal,* and is collected in large quantities. It is mostly shipped to Bordeaux, being largely consumed in France. The other varieties are derived from various species of *Acacia,* and are shipped from various ports, but by far the greater proportion of gum imported into this country comes from Egypt. The common name of *Gum Arabic,* which is applied to the various kinds of gum acacia, is a misnomer, as none is exported from Arabia.

General Characters and Tests.—The different varieties of gum acacia vary somewhat in their characters, but they may be described generally as follows :—In roundish, ovoid, or vermicular tears, or masses of various sizes ; or in angular fragments with glistening surfaces. When entire the tears or masses are either opaque from numerous fissures and brittle, or more or less transparent and not readily broken, the fractured surfaces being vitreous in appearance. They are colourless or nearly so, or of various shades of yellow, brown, or red, or of intermediate tints. They have no odour, and when of fine quality have a bland and mucilaginous taste. As stated in the pharmacopœia, gum acacia is also 'insoluble in alcohol, but entirely soluble in water, and forming a clear mucilaginous solution. The aqueous solution forms with subacetate of lead an opaque white jelly. If an aqueous solution of iodine be added to the powder, or to a solution formed with boiling water and cooled, there is no appearance of a violet or blue colour.'

Principal Constituents.—Gum acacia is essentially composed of *arabic* or *gummic acid,* combined with *calcium, magnesium,* and *potassium.*

Medicinal Properties.—Demulcent and emollient ; but its chief use is as a vehicle for the exhibition of other medicines.

Official Preparations.

Mistura Cretæ.	Pulvis Amygdalæ Compositus.
Mistura Guaiaci.	Pulvis Tragacanthæ Compositus.
Mucilago Acaciæ.	Trochisci, in all.

Pulvis Amygdalæ Compositus is also used in the preparation of Mistura Amygdalæ ; and Mucilago Acaciæ in all the Trochisci except Trochisci Opii.

---

19. ACACIA CATECHU, *Willd.*

Black Catechu.

(Bentley and Trimen's ' Medicinal Plants,' vol. ii. plate 95.)

Habitat.—This is a common tree in most parts of India and Burmah. It is also found in Ceylon, but is probably not known out of Asia.

Product Used and Name.—CATECHU NIGRUM :—an extract prepared principally from the wood of Acacia Catechu, *Willd.* ; and also, in part, from Acacia Suma, *Kurz.*

(*Not Official.*)

Catechu Nigrum.

Black Catechu.   Cutch.

Preparation and Commerce.—The mode of preparation of catechu varies somewhat in different localities, but the ordinary kind known in Europe, which is commonly termed *Pegu Catechu*, is an extract made from the wood of the two catechu trees mentioned above, or, according to some of their heart-wood only.

It is imported from British India, and is the produce of Bengal and Burmah, the best catechu coming from Pegu.

General Characters.— *Pegu Catechu, Cutch,* or *Black*

Catechu, is found in masses of irregular form, and commonly weighing several pounds, sometimes even as much as a hundredweight ; and made up of somewhat oblong pieces of varying length, breadth,and depth, enveloped in fragments of leaves.   It is dark rusty-brown or blackish in colour ; hard and brittle, and its fractured surface has a shining bubbly appearance.   It has no odour, but a very astringent and slightly bitter taste at first, succeeded by a sensation of sweetness. It is only partially soluble in cold water,—one of its constituents, catechin, b einnearly insoluble in that men-struum; but it is entirely soluble in alcohol and boiling water.

Principal Constituents.—Catechu-tannic acid and *catechin* ; the former a brown substance, readily soluble in water ; the latter (*catechin*) being in small acicular colourless crystals, and almost insoluble in cold water, as already noticed above.

Medicinal Properties.—Astringent like that of the official catechu from *Uncaria Gambier,* Roxb.

ORDER 2.—ROSACEÆ.

1. PRUNUS DOMESTICA, *Linn.,* var. JULIANA, *DC.*

St. Julien Plum.   French Plum.

(Bentley and Trimen's ' Medicinal Plants,' vol. ii. plate 96.)

Habitat. —This particular variety of *Prunus domestica* is largely cultivated in the valley of the Loire in France.

Official Part and Name.— PRUNUM :—the dried drupe. Imported from the South of France.

Prunum.

Prune.   St. Julien Plum.

Preparation and Commerce.   These fruits are dried partly by solar, and partly by artificial heat, being exposed on alternate days to the heat of an oven and to that of the sun. They are directed to be obtained from the South of France, the best coming from Bordeaux.

General Characters.—Prunes are somewhat ovoid or oblong in form, over an inch in length, black in colour, and more or less shrivelled. They have no very marked odour, but a sweet and somewhat mucilaginous acidulous taste. The pulp (*sarcocarp*), in which the properties of the fruit reside, has a brownish colour, and somewhat tough texture.

Principal Constituents.—There is no known constituent to account for its laxative properties; but its chief constituents appear to be *sugar, malic acid,* and *pectic* and *albuminoid substances.*

Medicinal Properties.—Demulcent and laxative.

Official Preparation.—Confectio Sennæ.

----

2. PRUNUS LAUROCERASUS, *Linn.*

Cherry Laurel.

(Bentley and Trimen's ' Medicinal Plants,' vol. ii. plate 98.)

Habitat.—This well-known plant is a native of the Caucasian provinces of Russia, and of other parts of the East; it has been introduced into, and now flourishes throughout, temperate Europe.

Official Part and Name.--LAUROCERASI FOLIA :—the fresh leaves.

Laurocerasi Folia.

Cherry Laurel Leaves.

General Characters.—The fresh leaves, which are alone official, are thick, coriaceous, on short stout petioles, oblong or somewhat obovate in outline, five to seven inches long by about one and a half to two inches broad, tapering towards each end, recurved at the apex, and distantly but sharply serrated and slightly revolute at the margins. The upper surface is dark green, smooth, and shining; the lower surface is dull, much paler, and with a very prominent midrib, on either side of which, towards the base, are one or two small yellowish or brownish rounded glandular depressions. When chewed the leaves have a bitter aromatic and somewhat astringent taste; but no odour, except when bruised or torn, when they emit a ratafia or bitter-almond-like odour.

Principal Constituents.—When the leaves are bruised or cut and submitted to distillation with water, they yield *hydrocyanic acid* and a volatile oil, which is regarded as identical with the *volatile oil of bitter almonds.* It is said to be produced by the decomposition of a nitrogenous principle contained in the leaves, which is analogous to, or probably a compound of, amygdalin, termed *laurocerasin*, under the influence of *synaptase* or *emulsin.*

Medicinal Properties.—Cherry-laurel leaves are usually administered in the form of the official cherry-laurel water, which possesses sedative properties owing to the presence of hydrocyanic acid.

Official Preparation.—Aqua Laurocerasi. *Dose.*—$\frac{1}{2}$ to 2 fluid drachms.

---

3. PRUNUS AMYGDALUS, *Stokes*
Almond.

Synonym.—AMYGDALUS COMMUNIS, *Linn.*

(Bentley and Trimen's ' Medicinal Plants,' vol. ii. plate 99.)

Habitat.—This tree is said to be a native of Morocco, Syria, Persia, and Turkestan; and doubtfully wild in Sicily, Greece, and Anatolia. It is cultivated in England, and throughout temperate Europe.

Official Parts or Products and Names.—1. AMYGDALA

DULCIS:—the ripe seed of the sweet almond tree, Prunus Amygdalus, *Stokes*, var. dulcis, *Baillon* (Amygdalus communis, *Linn.*, var. dulcis, *DC.*).   2. AMYGDALA AMARA :—the ripe seed of the bitter almond tree, Prunus Amygdalus, *Stokes*, var. amara, *Baillon* (Amygdalus communis, *Linn.*, var. amara, *DC.*).   3. OLEUM AMYGDALÆ :—the oil expressed from the bitter or sweet almond.

1. Amygdala Dulcis.

Sweet Almond.

General Characters and Varieties.—Sweet almonds are distinguished from bitter almonds by having a bland sweet agreeable nutty taste, and by their emulsion with water having no marked odour.   The varieties of sweet almonds now distinguished in the London market, and placed in the order of their value, are : *Jordan, Valencia, Sicily,* and *Barbary.*   Jordan almonds are the finest kind ; they are imported from Malaga, and *are alone official.*   Their characters are as follows :—About an inch or somewhat more in length, nearly oblong in form, more or less compressed, pointed at one end and rounded at the other, and covered by a cinnamon-brown scurfy coat or testa.   When macerated in warm water, this testa and the thin closely-adherent inner membrane or endopleura are removed, and the kernel, which is entirely formed of the embryo, alone remains, and is commonly known as the *blanched almond.*

The other varieties of the sweet almonds are distinguished from Jordan almonds by being shorter and somewhat ovoid in form, resembling in these respects bitter almonds, but readily distinguished from them by their taste and by the odour of their emulsion.   (*See* Amygdala Amara.)   From their greater length Jordan almonds are sometimes termed *long almonds.*

Almonds are sometimes imported in the shell (*endocarp*), in which case they are known as *almonds in the shell.*

Principal Constituents.—The *official fixed oil* is the principal constituent. (*See* Oleum Amygdalæ.) Sweet almonds also contain *sugar*, and two albuminoids, called *synaptase* or *emulsin* and *amandin*.

Medicinal Properties. —Sweet almonds are emollient and demulcent.

Official Preparations.

Oleum Amygdalæ (*which see*).

Pulvis Amygdalæ Compositus, which is also used for the preparation of Mistura Amygdalæ.

2. Amygdala Amara.
Bitter Almond

General Characters and Varieties.—Bitter almonds, as already noticed, resemble in form and length the inferior varieties of the sweet almond, and are hence readily distinguished like them from the official Jordan or long almond. They are also known from all kinds of sweet almonds by their very bitter taste, and by their aqueous emulsion having an odour like that of ratafia or peach-blossoms. They are distinguished, in the order of their value, as *French*, *Sicilian*, and *Barbary*.

Principal Constituents.—They yield by expression a similar *fixed oil* to that of sweet almonds, in the proportion of about 44 per cent. They also contain *sugar*, *emulsin*, and *amandin;* and likewise a nitrogenous crystalline glucoside called *amygdalin*, which is not present, at least in any appreciable proportion, in sweet almonds. The presence of amygdalin is, therefore, the special characteristic of bitter almonds; for when they are triturated with water an emulsion is formed having an odour like that of ratafia or peach-blossoms, which is caused by the decomposition of amygdalin under the influence of emulsin and water (the emulsin acting as a kind of ferment) into *glucose*, *hydrocyanic acid*, and *volatile oil of bitter almonds*. These latter substances are not, therefore, products of the bitter almond, but educts.

Medicinal Properties.—Sedative, owing to the formation of hydrocyanic acid as already described. They are too uncertain, however, to be employed internally, but the emulsion is sometimes used as a lotion in certain skin diseases.

Bitter almonds are sometimes used for flavouring, scenting, &c. But the volatile oil is more especially in use for such purposes; when employed, however, for flavouring, the purified oil of bitter almonds, or that which has been freed from hydrocyanic acid, should alone be used, as the crude oil is very poisonous.

Official Preparation.—Oleum Amygdalæ (*which see*).

3. Oleum Amygdalæ.

Almond Oil.

Extraction.—This oil may be expressed, as officially stated, from either the bitter or sweet almond, but it is almost exclusively obtained from the former, on account of their less cost and the greater value of the residual cake, which is used when mixed with water for the distillation of the volatile oil of bitter almonds. The produce of oil is, however, somewhat less from bitter than sweet almonds, for while the latter yield commonly about 50 per cent. of oil, the former, according to Umney, only yield about 44 per cent. In characters and composition the oil is in both cases the same.

General Characters.—Thin, pale yellow, nearly inodorous, and with a bland oleaginous taste. Its specific gravity is said to vary from 0·914 to 0·92; and it congeals about − 20° C. (− 4° F.). By exposure to air it soon becomes rancid.

Medicinal Properties.—Almond oil when applied locally is emollient; and when given internally it is laxative.

Official Preparations.

Oleum Phosphoratum.	Unguentum Simplex, and the
Unguentum Cetacei.	preparations containing it.
Unguentum Resinæ.	

4. HAGENIA ABYSSINICA, *Willd.*

Kousso.　Cusso.

Synonym.—BRAYERA ANTHELMINTICA, *Kunth.*

(Bentley and Trimen's 'Medicinal Plants,' vol. ii. plate 102.)

Habitat.—It is only found in Abyssinia.

Official Part and Name.—CUSSO:—the dried panicles (chiefly of the female flowers).

Cusso.

Kousso.

Collection and Commerce.—The panicles are unisexual; and both those of the male and female flowers, but chiefly the latter (*fig.* 23, B), are collected, and then suspended in the sun to dry; after which they are packed in boxes, and sent from Abyssinia to England by way of Aden or Bombay.

General Characters.—Kousso occurs in somewhat compressed clusters, which are either entire or more or less broken; or in sub-cylindrical rolls bound together by transverse bands; or, occasionally, the panicles are broken up into small fragments. The clusters or rolls vary in length from ten inches to one foot or more; they are brownish or greenish-brown, or reddish in the case of the female flowers (*red kousso*); they have a pleasant herby tea-like odour; and a bitter acrid disagreeable taste. The separate panicles are much branched, zigzag (*fig.* 23, A and B), more or less covered with hairs and glands, and with a large sheathing bract at the base of each branch. The flowers (*fig.* 23 B

and c) are numerous, small, shortly stalked, unisexual,
with two rounded membranous veiny bracts at the base of

FIG. 23.—A. Branch with leaves and flowers of *Hagenia abyssinica*. B. Portion
of female panicle.   c. Female flowers seen laterally.   D. A single female flower.
a, b, c, d, e. The five outer segments (epicalyx) of the calyx.

each flower, which are greenish-yellow in the male, and
tinged with red in the female flowers.   The calyx is hairy

L

externally and veiny, with ten segments in two alternating
whorls (*fig.* 23, D, *a*, *b*, *c*, *d*, *e*). There are five small linear-
lanceolate petals, which are inserted into the throat of the
calyx. The stamens in the male flowers vary from fifteen
to thirty in number, and are inserted into the throat of
the calyx-tube; and the female flowers have usually two
carpels enclosed in the deep calyx-tube, and each surmounted by a style, which projects from the tube. The
fruit, when present, is an ovoid membranous one-seeded
achene.

Principal Constituents.—The active principle is said to
be *koussin* or *kosin.* But, according to Buchheim, koussin,
when pure, is almost inert. Other constituents of kousso
are a *volatile oil, bitter acrid resin,* and *tannic acid.*

Medicinal Properties.—Anthelmintic. It is said to be
effectual in destroying both kinds of tapeworm, namely, the
Tænia solium and *Bothriocephalus latus.*

Official Preparation.—Infusum Cusso.

5. ROSA CANINA, *Linn.*
Dog Rose.

(Bentley and Trimen's ' Medicinal Plants,' vol. ii. plate 103.)

Habitat.—This is a common plant in Great Britain and
throughout Europe. It extends eastward into Northern
Asia and Persia; and also occurs in North Africa and the
Canaries.

Official Part and Name.—ROSÆ CANINÆ FRUCTUS :—
the ripe fruit of Rosa canina, *Linn.*, and other indigenous
allied species.

Rosæ Caninæ Fructus.

Fruit of the Dog Rose. Hips.

General Characters.—Hips are three-quarters of an inch or more in length, ovoid or somewhat oval in form; smooth, shining, and of a scarlet or red colour externally. The part within the external skin is soft, pulpy, of an orange colour, and inodorous; but with a pleasant sweetish acidulous taste. Hips consist essentially of an enlarged fleshy concave thalamus and adherent calyx-tube, which is crowned with the remains of its segments, and enclosing numerous hard brown hairy achenes; the fruit thus formed is termed a *cynarrhodum.* The only part used in medicine is the pulp.

Principal Constituents.—*Malic acid, citric acid, sugar,* and *gum.*

Medicinal Properties.—Refrigerant and mildly astringent; but essentially used in the form of the official confection as an excipient.

Official Preparation.—Confectio Rosæ Caninæ.

---

6. ROSA GALLICA, *Linn.*

Red Rose. French Rose.

(Bentley and Trimen's ' Medicinal Plants,' vol. ii. plate 104.)

Habitat.—Wild or semi-wild throughout Europe, except the northern parts (which include the British Islands and Scandinavia), and extending to Greece, the Crimea, and Armenia.

Official Part and Name.—ROSÆ GALLICÆ PETALA :— the fresh and dried unexpanded petals.

Rosæ Gallicæ Petala.

Red Rose Petals.

Collection and Preparation.—The official *red rose petals*, or, as they are often incorrectly termed, *red rose leaves*, are directed to be obtained from plants cultivated in Britain. They are collected and dried as follows:—The flower-buds are gathered just before their expansion; the petals are then cut off, so as to leave the narrow whitish portions near the base, attached to the calyx. They are then carefully dried by stove heat; and after being sifted, so as to remove the stamens and other extraneous substances, are ready for the market.

The petals are also used in a fresh state for making Confection of Roses.

General Characters.—As found in the pharmacies, they are usually in loosely aggregated cone-like masses; or sometimes they are separate, and more or less crumpled. The petals, when well preserved, have a fine purplish-red colour, velvety appearance, are crisp and dry, have a fragrant roseate odour, and a bitterish, feebly acid, and astringent taste.

Principal Constituents.—A trace of *volatile oil, red colouring matter*, a little *gallic* and *quercitannic acids*, and *glucose*.

Medicinal Properties. — Slightly astringent and tonic. The infusion and confection are chiefly used for the exhibition of other more active medicines.

Official Preparations.

1. OF ROSÆ GALLICÆ PETALA (fresh):—

Confectio Rosæ Gallicæ, which is also used in the preparation of several of the official Pilulæ.

2. OF ROSÆ GALLICÆ PETALA (dried):—

Infusum Rosæ Acidum. | Syrupus Rosæ Gallicæ.

7. ROSA CENTIFOLIA, *Linn.*
Cabbage Rose.

(Bentley and Trimen's 'Medicinal Plants,' vol. ii. plate 105.)

Habitat.—This rose is said to be a native of Persia, Assyria, and of the eastern part of the Caucasus, where it is found with single flowers. It occurs in a cultivated state, and with more or less double flowers, in innumerable varieties in all temperate regions.

Official Part and Name.—ROSÆ CENTIFOLIÆ PETALA : the fresh fully-expanded petals.

Rosæ Centifoliæ Petala.
Cabbage Rose Petals.

Collection and Preparation.—Cabbage rose petals, or petals of the *Hundred-leaved rose* as they are also termed, are directed to be obtained from plants cultivated in Britain; and, as their odour is most fragrant when the flowers are full-blown, they are ordered to be used in this state. When dried, their fragrance is diminished, and they become of a brown colour. They may, however, be preserved for some time by intimately mixing them with common salt, and pressing the mixture into a closely-stopped vessel, and kept in a cool place.

General Characters.—The fresh petals are large, thin, delicate, beautiful pink or white, deliciously fragrant, and with a sweetish, slightly astringent, bitterish taste. Both taste and odour are readily imparted to water.

Principal Constituents.—Very similar to those of red-rose petals. Their odour is due to the presence of a very small quantity of a *volatile oil.* This oil has a faint rose-like, but scarcely agreeable odour, and a butyraceous consistence. It is principally composed of an inodorous *stearoptene.* This oil is of no value as a perfume, like the deliciously fragrant *oil, attar,* or *otto of Rose,* which is obtained by distilling with water the fresh flowers or petals of *Rosa damascena,* Miller.

Medicinal Properties.—Mildly laxative; but scarcely used at the present day, except in the preparation of the official Rose water.

Official Preparation.—Aqua Rosæ.

Aqua Rosæ is also used in the preparation of Mistura Ferri Composita and Trochisci Bismuthi.

Order 3.—HAMAMELIDACEÆ.

1. LIQUIDAMBAR ORIENTALIS, *Miller.*
Oriental Liquidambar.

(Bentley and Trimen's ' Medicinal Plants,' vol. ii. plate 107.)

Habitat.—This tree appears to be confined to a very restricted portion of Asia Minor, on and near the coast in the south-west.

Official Product and Name.—STYRAX PRÆPARATUS :—a balsam prepared from the inner bark.   Purified by solution in spirit, filtration, and evaporation.

Styrax Præparatus.
Prepared Storax.

Extraction and Commerce. — The balsam known as *Liquid Storax* is either obtained from the inner bark by first submitting it to pressure, then throwing hot water upon it, and subsequent pressure ; or by first boiling the bark in water and skimming off the balsam, and subsequently pressing the boiled bark, and mixing the products.   It is then forwarded in barrels, etc., to Smyrna, Constantinople, and other ports, and reaches England chiefly by way of Trieste.

The bark left after the balsam has been extracted, constitutes the fragrant foliaceous cakes known as *Cortex
Thymiamatis* or *Storax Bark.*

General Characters and Tests.—Liquid Storax as imported
is a viscid, opaque, semi-liquid balsam, of a greyish-brown
colour. It always contains a variable proportion of water,
which, after it has been allowed to stand for some time, floats
on the surface. It is also frequently contaminated by the
admixture of sand, ashes, fragments of bark, etc., hence the
official storax is directed to be prepared from liquid storax
by solution in spirit, filtration, and evaporation. It is then
called Prepared Storax, and its characters and tests are given
as follows :—A semi-transparent brownish-yellow semi-fluid
balsam about the consistence of thick honey, with a strong
agreeable odour and balsamic taste. Heated in a test-tube
on the vapour-bath, it becomes more liquid, but gives off no
moisture ; boiled with solution of bichromate of potassium
and sulphuric acid, it evolves an odour resembling that of
essential oil of bitter almonds.

Principal Constituents.—Its most important constituent
is *styrol, cinnamol,* or *cinnamene,* which is a colourless liquid
volatile hydrocarbon, having the odour and burning taste
of the balsam. Other constituents are, *cinnamic acid,* the
amorphous substance termed *storesin,* and *styracin* or *cinnamate of cinnamyl* in crystals.

Medicinal Properties.—Stimulant and expectorant, like
benzoin and the balsams of tolu and peru, but very little
used at the present day. Its chief use is probably as a local
application mixed with oil, etc.

Official Preparation.—Tinctura Benzoini Composita.

ORDER 4.—MYRTACEÆ.

1. MELALEUCA MINOR, *Sm.*
Cajuput.

Synonym.—MELALEUCA CAJUPUTI, *Roxb.*

(Bentley and Trimen's ' Medicinal Plants,' vol. ii. plate 108.)

Habitat.—This species, according to Bentham, occurs in several forms, and is widely distributed; but that form from which Cajuput oil is obtained grows only in several of the East Indian Islands, more especially in Celebes, Bouro, and Amboyna. It may also probably occur in the Philippines, Cochin China, and New Caledonia.

Official Product and Name.—OLEUM CAJUPUTI :—the oil distilled from the leaves.

Oleum Cajuputi.
Oil of Cajuput.

Extraction and Commerce.—The oil is obtained by distilling the leaves with water in the most primitive manner. It is chiefly derived from Celebes, but some is also obtained from Java, Manila, Bouro, and other places ; it is principally imported from Singapore and Batavia.

General Characters.—Oil of Cajuput or Cajeput is a transparent limpid very volatile liquid, of a pale bluish-green colour, a strong penetrating agreeable camphoraceous odour, and a warm bitterish aromatic camphoraceous taste, succeeded by a sensation of coldness in the mouth. Its average specific gravity is ·926, although it is said to vary from ·914 to ·930 ; and does not congeal at 8·6° Fahr. It is levogyre.

Principal Constituents.—Its chief constituent is *cajuputol* or *hydrate of cajuputene*, which is a hydrocarbon with an agreeable odour resembling hyacinths. The green colour of the oil, as found in commerce, is due to copper, which may be detected in minute proportion in all that is imported. The natural green colour of the oil is soon lost by keeping, hence it is supposed to be contaminated with copper in order to make the green tint permanent.

Medicinal Properties. — Administered internally, it is stimulant, antispasmodic, and diaphoretic; and externally applied it is stimulant and rubefacient. *Dose.*—1 to 4 minims.

Official Preparations.

Linimentum Crotonis. | Spiritus Cajuputi.

---

2. EUCALYPTUS GLOBULUS, *Labill.*

Blue Gum Tree.

(Bentley and Trimen's ' Medicinal Plants,' vol. ii. plate 109.)

Habitat.—This fine tree is a native of the southern half of Tasmania. It is also found in Flinders Island, and on the mainland of Australia in Victoria.

Official Product and Name.—OLEUM EUCALYPTI :—the oil distilled from the fresh leaves of Eucalyptus Globulus, *Labill.* ; Eucalyptus amygdalina, *Labill.* ; and probably other species of Eucalyptus. (The best oil is now said to be obtained from *E. dumosa.*)

Oleum Eucalypti.

Oil of Eucalyptus.

Commerce.—This oil is imported to a large extent from Australia.

General Characters.—Oil of Eucalyptus is colourless or pale straw-yellow, becoming darker and thicker by exposure. It has an aromatic odour, and a pungent spicy taste, leaving a sensation of coldness in the mouth. It is neutral to litmus paper. Specific gravity averages nearly ·900. It is soluble in about an equal weight of alcohol. The above characters are those given in the pharmacopœia, but the oils of commerce are now found to vary much as regards their specific gravity, in their relations to tests, and in other characters, according to their botanical source. The oil from *Eucalyptus Globulus* has been found to be dextrogyre, and this is probably always the case with eucalyptus oils.

Principal Constituents.—The essential constituent, according to Cloëz, is *eucalyptol*, which is now said to be a mixture of one or two *terpenes* and *cymene* or *cymol.*

Medicinal Properties.—Externally applied it is rubefacient and powerfully antiseptic. Administered internally it is antipyretic and antiperiodic. *Dose.*—1 to 4 minims.

Official Preparation.—Unguentum Eucalypti.

3. PIMENTA OFFICINALIS, *Lindl.*
Pimento.

Synonym.—EUGENIA PIMENTA, *DC.*

(Bentley and Trimen's 'Medicinal Plants,' vol. ii. plate 111.)

Habitat.—It is a native of most of the West Indian Islands; and is also found in Central America, and in Venezuela.

Official Parts or Products and Names.—1. PIMENTA:— the dried unripe full-grown fruit. 2. OLEUM PIMENTÆ :— the oil distilled in Britain from pimento.

1. Pimenta.
Pimento.

Collection, Preparation, and Commerce.—The fruits when fully ripe lose their aromatic properties, hence they are collected when full grown but while yet green. They are then dried by exposure to the sun and air for several days, after which they are separated from their pedicels and put into bags and casks for exportation. Pimento is entirely derived from Jamaica, where the trees are largely cultivated.

General Characters. — *Pimento, Allspice,* or *Jamaica Pepper,* is dry, light, roundish, one-fifth of an inch or more in diameter, crowned with the remains of the calyx in the form commonly of a slightly raised scar-like ring, surrounding a short depressed style ; and having rarely at the other end a short stalk. It consists of a thin brittle dark-brown somewhat woody shell or pericarp, which is roughish from the presence of projecting oil-glands ; and of two dark brownish-black somewhat compressed reniform seeds, each of which is contained in a separate cell. Its odour is agreeably aromatic, and has been thought to resemble a mixture of clove, cinnamon, and nutmeg, and hence its common name of allspice ; and it has a warm agreeable spicy taste, much resembling that of cloves.

Principal Constituents.—Its aromatic properties are due to a *volatile oil,* which is official (see *Oleum Pimentæ*). *Starch* is also an ingredient, and it is rich in *tannic acid.*

Medicinal Properties.—Aromatic, carminative, and stimulant, resembling cloves ; but its chief use is as a culinary spice.

Official Preparations.

Aqua Pimentæ. | Oleum Pimentæ.

2. Oleum Pimentæ.

Oil of Pimento.

Preparation and General Characters.—This oil, which is directed to be distilled in Britain, is most abundant in the pericarp, and is obtained from the whole fruit in the proportion of from about 3 to $4\frac{1}{2}$ per cent. It is colourless or slightly yellowish-red when recent, but becoming brown by age. Its odour and taste resemble those of pimento in a concentrated degree. It has a specific gravity of about 1·040, and sinks in water ; it is slightly acid and levogyrate. The oil is a mixture of a hydrocarbon, called *light oil of pimento,* and of an oxygenated oil, termed *eugenol, eugenic acid,* or *heavy oil of pimento.*

Medicinal Properties.—Similar to those of pimento, for internal use ; but it is also employed externally. *Dose.*—1 to 4 minims.

4. EUGENIA CARYOPHYLLATA, *Thunb.*
Clove.

Synonym.—CARYOPHYLLUS AROMATICUS, *Linn.*

(Bentley and Trimen's ' Medicinal Plants,' vol. ii. plate 112.)

Habitat.—This tree is supposed to have been originally a native only of the five small islands forming the true Moluccas or Clove Islands. It is not now found in these islands, but is largely cultivated in the island of Amboyna, also at Sumatra and Penang, in the islands of Zanzibar and Pemba, and in the West Indies and other tropical regions.

Official Parts or Products and Names.—1. CARYOPHYLLUM :—the dried flower-bud. 2. OLEUM CARYOPHYLLI :—the oil distilled in Britain from cloves.

1. Caryophyllum.
Clove.

Collection, Preparation, Commerce, and Varieties.—The mode of collection and preparation of cloves, although varying in certain particulars, essentially consists in gathering the flower-buds either entirely by hand, or partly by hand and partly by beating them off the trees by bamboos on to cloths placed beneath them, as soon as they have lost their green colour and become red; they are then simply dried by exposure in the sun by which they acquire their ordinary brown colour. Cloves are chiefly obtained from the islands of Zanzibar and Pemba, and from the Molucca Islands. There are several varieties of cloves; but those most esteemed in this country are Penang, Bencoolen, Amboyna, and Zanzibar, which are arranged in the order of their value.

General Characters.—Cloves are usually somewhat over half an inch in length, and consisting of a long dark brown sub-cylindrical and somewhat angular wrinkled calyx-tube, which tapers below, and is surmounted by four teeth forming the limb of the calyx, between which the four paler-coloured petals, enclosing the numerous stamens and style, are rolled up in the form of a ball. They have a strong fragrant spicy odour; and a very pungent and aromatic taste. Good cloves are large, plump, heavy, dark brown in colour, and when slightly pressed or scraped with the nail a small quantity of oil exudes from them. Inferior kinds are smaller, paler-coloured, light in weight, more or less shrivelled, and from being less rich in oil they have a feebler taste and odour.

Principal Constituents.—Cloves owe their properties to a *volatile oil*, which is official, and is described below. Other constituents are *gum*, *tannic acid*, and a colourless tasteless substance, which crystallises in needle-shaped prisms, called *caryophyllin*, a little *salicylic acid*, &c.

Medicinal Properties.—Cloves are stimulant, aromatic, and carminative, and although much used in combination with other medicines, their chief consumption is as a culinary spice.

Official Preparations.

Infusum Aurantii Compositum.
Infusum Caryophylli.
Mistura Ferri Aromatica.

Oleum Caryophylli.
Pulvis Cretæ Aromaticus.
Vinum Opii.

2. Oleum Caryophylli.

Oil of Cloves.

Preparation and General Characters.—This oil is directed to be distilled in Britain from cloves. Cloves are remarkable for the large amount of oil they yield,—that is, from about 17 to 20 per cent. Oil of cloves is colourless or pale yellow when recent, but gradually acquires a reddish-brown tint. It has in a high degree the odour and taste of cloves. Its specific gravity varies from about 1·046 to 1·058 ; it sinks in water. It is slightly acid, and has no sensible rotatory power. Like oil of pimento, it is a mixture of a light oil or hydrocarbon, termed *light oil of cloves*, and of an oxygenated oil called *heavy oil of cloves, eugenol,* or *eugenic acid,* which has the full taste and odour of cloves.

Medicinal Properties.—Similar to cloves for internal administration ; but it is also used externally. *Dose.*—1 to 4 minims.

Official Preparations.

Confectio Scammonii. | Pilula Colocynthidis Composita.
Pilula Colocynthidis et Hyoscyami.

5. PUNICA GRANATUM, *Linn.*

Pomegranate.

Bentley and Trimen's ' Medicinal Plants,' vol. ii. plate 113.)

Habitat.—It is regarded as indigenous in North-Western India, Southern Persia, and perhaps Palestine. It has been long cultivated, and is now spread over all the warmer and temperate countries of the globe.

Official Part and Name.—GRANATI RADICIS CORTEX :— the dried bark of the root.

1. Granati Radicis Cortex.

Pomegranate Root Bark.

Commerce.—It is derived from the South of Europe, and is more especially collected in the South of France and in Italy, and imported in a dried state.

General Characters.—This bark is found in small quills or fragments, varying commonly from two to four inches in length, and being about one-twenty-fifth of an inch or somewhat more in thickness. The outer surface is yellowish-grey, wrinkled or marked with faint longitudinal striæ, or more or less furrowed with bands or conchoidal scales of cork ; its inner surface is smooth or finely striated, yellowish, and frequently having portions of the pale-coloured wood adhering to it. The fracture is short and somewhat granular ; it has no odour, but a marked astringent and very feebly bitter taste.

Principal Constituents.—It contains over 22 per cent. of a variety of tannic acid called *punico-tannic acid*, much *mannite*, and about one-half per cent. of a colourless liquid somewhat aromatic alkaloid, named *pelletierine*, which is soluble in water, alcohol, and chloroform, and is regarded by Tanret, its discoverer, as the source of its anthelmintic property.

Adulterations and Substitutions.—Box bark from *Buxus sempervirens*, and Barberry bark from *Berberis vulgaris*, are sometimes mixed with, or substituted for, pomegranate root bark, but are readily distinguished by their infusions not being affected by the addition of a persalt of iron, whereas an infusion of pomegranate root bark becomes deep blackish-blue under similar circumstances. These barks have also a marked bitter taste.

Pomegranate stem bark, which is commonly regarded as less active than the root bark, but on no good authority, is also frequently mixed with, or substituted for it. Its less corky character is the best characteristic, for while it has longitudinal bands or ridges of cork, it has no broad flat scales.

Medicinal Properties.—Astringent, and a valuable anthelmintic for tapeworm. The sulphate and tannate of pelletierine are also now recommended as effective tænicides.

Official Preparation.—Decoctum Granati Radicis.

2. Granati Fructus Cortex.

Pomegranate Rind or Peel.

(Not Official.)

General Characters.—The dried rind or pericarp occurs in more or less arched irregular hard brittle fragments, some of which have the tubular toothed calyx with the remains of the stamens and style projecting from them. The fragments are yellowish- or reddish-brown, and somewhat rough externally ; and on their inner surface they are lighter coloured, being pale yellow or brownish, and marked with depressions left by the seeds. The rind has no marked odour, but a very astringent taste.

Principal Constituents.-- Like the root bark it contains a large proportion of *tannic acid*, and hence its infusion produces with a persalt of iron an abundant blackish-blue precipitate.

Medicinal Properties.—Astringent, but its use is now almost obsolete in this country.

ORDER 5.—CUCURBITACEÆ.

1. CITRULLUS COLOCYNTHIS, *Schrad.*

Colocynth.

Synonym.—CUCUMIS COLOCYNTHIS, *Linn.*

(Bentley and Trimen's ' Medicinal Plants,' vol. ii. plate 114.)

Habitat.—This plant has a wide distribution, extending from North-West India, throughout Arabia and Syria to

Egypt, where it is very abundant, some of the Greek Islands, Northern Africa to Morocco, and the Cape de Verd Islands. It is also found in the South-East of Spain and Portugal, at the Cape of Good Hope, and in Japan.

Official Part and Name.—COLOCYNTHIDIS PULPA :—the dried peeled fruit, freed from seeds.

Colocynthidis Pulpa.

Colocynth Pulp.

Collection and Preparation.—The fruit is gathered in the autumn, and is then either peeled when fresh and dried quickly in a stove, or more slowly by exposure to the sun; or it is peeled after being dried. Rarely the dried unpeeled fruit is imported, but such colocynth is not official.

Varieties and Commerce.—Two kinds of colocynth have been distinguished, namely, *Turkey* or *Peeled Colocynth*, and *Mogador* or *Unpeeled Colocynth.* The former is commonly imported from Spain and Syria, and is the official kind described in the pharmacopœia; the latter is obtained from Mogador.

General Characters and Tests.—The dried peeled fruit (pepo), as imported, is usually in more or less broken balls consisting of the pulp in which the seeds are imbedded. These balls are two inches or less in diameter, roundish, somewhat angular on their surface, and frequently with small pieces of the rind at the angles, nearly white or with a tinge of yellow, very light in weight, spongy, tough, and easily breaking up vertically into three wedge-shaped portions, in each of which, near the rounded surface, numerous flat ovoid brownish or yellowish seeds are placed. The broken-up pulp freed from seeds is the condition in which it is usually supplied to pharmacists under the name of *colocynth pulp* or *pith.* This pulp, which is *alone official*, is light in weight, spongy, tough, whitish, inodorous, but with an intensely bitter taste. The powder is not coloured blue by iodine;

M

and does not yield oil when treated with ether, and the separated ether evaporated.

The variety known as Mogador colocynth is larger than the Turkey variety, being three or more inches in diameter. It is covered by a smooth yellowish-brown firm rind, hence its common name of *unpeeled colocynth.* It is inferior in quality to the so-called Turkey colocynth, and should not, therefore, be substituted for it.

Principal Constituents.—The active principle of colocynth pulp is a glucoside termed *colocynthin,* which is resolvable into *glucose* and a resinous substance called *colocynthein.* Colocynthin has been obtained both in an amorphous and crystalline condition ; it is yellow, very bitter, and soluble in both water and alcohol. Starch is not a constituent of the pulp, hence the powder is not coloured blue by iodine.

Medicinal Properties.—A powerful hydragogue cathartic, and also diuretic. The active principle, colocynthin, has similar properties, and has occasionally been employed medicinally, but its action is so powerful that its use is not to be recommended.

Official Preparations.

Extractum Colocynthidis Compositum.
Pilula Colocynthidis Composita.
Pilula Colocynthidis et Hyoscyami.

---

2. ECBALLIUM ELATERIUM, *A. Rich.*

Squirting Cucumber.

Synonym.—MOMORDICA ELATERIUM, *Linn.*

(Bentley and Trimen's ' Medicinal Plants,' vol. ii. plate 115.)

Habitat.—A common weed in the South of Europe throughout the Mediterranean district, and reaching as far to the east as Persia.

Official Parts or Products and Names.—1. ECBALLII

FRUCTUS :—the fruit, very nearly ripe.   2. ELATERIUM :—a sediment from the juice of the Squirting Cucumber fruit. 3. ELATERINUM :—the active principle of elaterium.

1. Ecballii Fructus.
Squirting Cucumber Fruit.
Synonym.—Elaterii Fructus.

Collection.—The fruit is directed in the British Pharma-copœia to be obtained from plants cultivated in Britain, and when very near ripe.   The latter direc-tion is necessary, for if the fruit is left till it is quite ripe, it separates spontaneously from its peduncle (*fig.* 24, *c*), and the seeds and juice are at the same moment forcibly expelled (*fig.* 24, *a*) from the orifice thus formed by the de-tached peduncle ; hence the common name of Squirting Cucumber which is applied to it. This singular action is due to osmosis, by means of which the juice of the outer part of the pericarp gradually passes through its contractile inner portion, until at length the central pulp becomes so engorged,

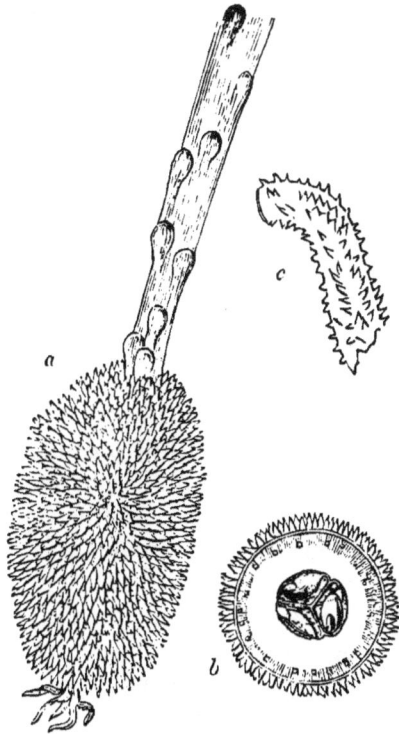

FIG. 24.—*a.* Fruit (*pepo*) of the Squirting Cucumber (*Ecballium Elaterium*) dischar-ging its seeds and juice. *b.* Transverse section of the fruit. *c.* The separated or broken-off stalk (*peduncle*).

and the tension so great, that the wall gives way at its

M 2

weakest point, which is where the fruit is attached to its stalk.

General Characters.—The fruit is pendulous from the recurved apex of a succulent tapering slightly hispid peduncle. It is oblong-ovoid in form, rounded at the base, and crowned at the apex by the withered remains of the flower; from one and a half to more than two inches long, firm, yellowish-green, and covered with short pale fleshy processes terminating in hair-like white points (*fig.* 24, *a*). The pericarp is thick, fleshy, white within, and filled with a watery juice, in which numerous somewhat oblong seeds are immersed (*fig.* 24, *b*).

2. Elaterium.

Elaterium.

Synonym.—Extractum Elaterii.

Preparation.—As the juice immediately surrounding the seeds is the most active part of the fruit, the finest elaterium is obtained from the juice which runs spontaneously from the sliced ripe fruit. But in practice some pressure must be applied, because, from the circumstances already mentioned (page 163), the fruits must be gathered before they are quite ripe, and in this condition the juice does not readily flow from them until lightly pressed. Hence the process for its preparation given in the pharmacopœia.

The amount of elaterium obtainable will vary much according to the amount of pressure, and also to some extent to the seasons. Thus in a very warm dry season the produce will be larger than in a cold wet one; but on an average the quantity of elaterium obtainable, when prepared according to the directions of the pharmacopœia, is about half an ounce from a bushel, or forty pounds, of the fruits.

General Characters and Tests.—In light friable flat or slightly curved opaque cakes, about one-tenth of an inch thick, and usually marked on one of their surfaces by the impression of the paper, linen, or muslin, on which they were dried. It is pale green, greyish-green, or yellowish-grey, according to its age ; has a finely granular fracture ; an acrid and bitter taste ; and a faint tea-like odour. It does not effervesce with acids ; boiled with water and the cooled mixture treated with iodine, affords little or no blue colour ; yields half its weight to boiling rectified spirit. Treated by the method described for obtaining elaterin in the British Pharmacopœia, it should yield 25 per cent., or not less than 20 per cent., of that substance.

Principal Constituents.—The active principle of elaterium is *elaterin*, which is official, and is, therefore, described below.

3. Elaterinum.

Elaterin.

This is best obtained according to the process of the British Pharmacopœia, and its characters and tests are given in that volume. It has been demonstrated that hot dry weather is favourable to the development of elaterin, and hence elaterium is most powerful when produced in very fine dry summers. From this circumstance, as also from the fact that elaterium varies in strength from climate, time, and mode of preparation, elaterin is a more definite and reliable substance than it, and has, consequently, been now made official.

Medicinal Properties.—Elaterium, and more especially its active principle elaterin, are the most powerful hydragogue cathartics known.

Official Preparations.

1. Of Ecballii Fructus :—
      Elaterium. *Dose.*—$\frac{1}{16}$ to $\frac{1}{2}$ a grain.
2. Of Elaterium :—
      Elaterinum. *Dose.*—$\frac{1}{40}$ to $\frac{1}{10}$ of a grain.
3. Of Elaterinum :—
      Pulvis Elaterini Compositus. *Dose.*—$\frac{1}{2}$ grain to 5 grains.

ORDER 6.—UMBELLIFERÆ.

1. CONIUM MACULATUM, *Linn.*
Hemlock.

(Bentley and Trimen's 'Medicinal Plants,' vol. ii. plate 118.)

Habitat.—It is more or less distributed throughout Britain; and also in temperate Europe and Asia. It also occurs in Asia Minor and the Mediterranean Islands.

Official Parts and Names.—1. CONII FOLIA :—the fresh leaves and young branches ; gathered from wild British plants when the fruit begins to form. 2. CONII FRUCTUS :—the fruit, gathered when fully developed, but while still green, and carefully dried.

1. Conii Folia.

Hemlock Leaves.

Collection.—The leaves and young branches should be gathered at the time directed in the pharmacopœia, for the reasons explained under 'Aconiti Folia.'

General Characters and Test.—The leaves are more or less divided in a pinnate manner, the lower ones decompound, deltoid in outline, and very large, even sometimes two feet in length ; all glabrous, dull dark green, and arising from a very smooth hollow striated bright green stem, which is mottled with small irregular dark purple spots or stains, by clasping petioles of varying lengths, those of the lower leaves being hollow. Odour strong and very disagreeable, more especially when rubbed with solution of potash.

Principal Constituents.—Hemlock leaves have essentially the same active constituents as the fruit, but in much smaller proportion. Their nature is described below under 'Conii Fructus.'

2. Conii Fructus.

Hemlock Fruit.

Collection and Preparation.—In the British Pharmacopœia of 1867 the dried ripe fruit was directed to be used, but the researches of Sir Robert Christison, Dr. Manlius Smith, and Dr. John Harley having demonstrated the greater activity of the full-grown fruit gathered while still green, it has been so ordered to be used in the present pharmacopœia. Harley states that the 'fruits contain the largest amount of conia just before they come to maturity, that is, when they have attained their full size, but are still soft and green. At this stage they should be collected, spread out in thin layers on porous paper, dried in a warm shady room, at about the temperature of 80°, and then preserved in a dry place in well-closed tin canisters. With these precautions they will retain their virtues unimpaired for more than a year.'

General Characters and Test.—Hemlock fruits (*fig.* 25, *a*) are about one-eighth of an inch long, broadly ovoid in form,

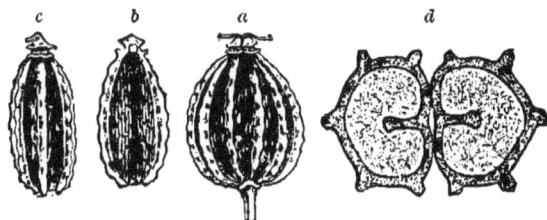

FIG. 25.—Conii Fructus, *Hemlock Fruit.* *a.* Entire fruit. *b.* Commissural surface of a mericarp. *c.* Dorsal surface of a mericarp. *d.* Transverse section of the fruit. (All enlarged.)

somewhat compressed laterally, and crowned by the depressed stylopod, and of a dull greenish-grey colour. The half-fruits (*fig.* 25, *b*, *c*) or mericarps, are commonly seen in the pharmacies mixed in variable proportions with the entire fruits; each of these presents five prominent more or less crenated or wavy ridges, with the furrows smooth or nearly so, and without evident vittæ, and when cut transversely, the contained seed, from having a deep furrow on its inner surface, presents a reniform outline (*fig.* 25, *d*). Hemlock fruits have but little taste, or even odour except when triturated with solution of potash, when it is very strong and disagreeable.

Principal Constituents.—The most important constituent of both leaves and fruits is *conine* or *conia* ; but in no case, it is said, do the fruits, which contain the most, yield more than about $\frac{1}{2}$ per cent. Conia is a volatile, colourless, oily, strongly alkaline liquid, with a very strong disagreeable odour, powerfully poisonous properties, and is freely soluble in ether and alcohol. It is upon its presence that the medicinal activity both of the fruit and leaves essentially depend. Two other alkaloids have also been found in hemlock fruits, namely, *methyl-conine*, to which it is probable some of their activity is due ; and a less poisonous base, named *conhydrine*.

Medicinal Properties.—Both leaves and fruits have sedative and antispasmodic properties; and in certain doses hemlock acts as a paralyser to the centres of motion. Conine has also been employed both internally by the stomach and hypodermically ; but according to Harley, it is variable both in strength and action, and has not proved a satisfactory drug.

Official Preparations.

1. Of CONII FOLIA :—Extractum Conii, which is also an ingredient in the preparation of Pilula Conii Composita ; and Succus Conii, which is also used in the preparation of Cataplasma Conii and Vapor Coninæ.

Dose of Extractum Conii, 2 to 6 grains. *Dose* of Succus Conii, $\frac{1}{2}$ to 1 fluid drachm.

2. Of CONII FRUCTUS :—Tinctura Conii. *Dose.*—20 to 60 minims.

2. CARUM CARUI, *Linn.*
Caraway.

(Bentley and Trimen's ' Medicinal Plants,' vol. ii. plate 121.)

Habitat.—Caraway is found in this country, and also distributed over the northern parts of Europe and Asia, but it is difficult in many cases to determine whether truly indigenous or originally introduced.

Official Parts or Products and Names. — 1. CARUI FRUCTUS:—the dried fruit. 2. OLEUM CARUI :—the oil distilled in Britain from caraway fruit.

1. Carui Fructus.
Caraway Fruit.

Varieties and Commerce.—There are several varieties of caraway fruit distinguished in commerce, but those best known in the London market are English, Dutch, and German, the first being regarded as the best. The first is derived from plants cultivated in Kent and Essex ; Dutch caraways are largely imported from Holland ; and German caraways are obtained from plants which are extensively cultivated in Moravia and Prussia. Other varieties are imported from Norway, Finland, Russia, and the ports of Morocco.

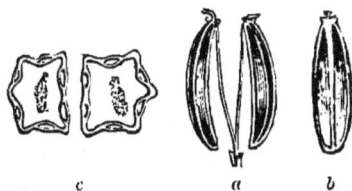

FIG. 26.—Carui Fructus, **Caraway Fruit.** *a.* Entire fruit. *b.* Commissural surface of a mericarp. *c.* Transverse section of the fruit. (All enlarged.)

General Characters.—As seen in commerce, caraway fruits, commonly known as *caraways*, are usually separated into their constituent mericarps, or rarely the two mericarps are loosely attached to the central axis or carpophore of the fruit (*fig.* 26, *a*). Each mericarp (*fig.* 26, *b*), which in common language is improperly designated a seed, varies in length in the different varieties, being from about one-sixth to one-fourth of an inch long ; excepting the Mogador kind, which is longer. It is slightly curved, somewhat tapering at each end, brown, and marked with five paler-coloured longitudinal ridges, and in each intervening space there is a large and conspicuous vitta, and each mericarp has also two vittæ on its face (*fig.* 26, *c*). Caraway fruit has an agreeable aromatic odour; and a pleasant sweetish spicy taste.

Principal Constituents.—The properties of caraway fruit depend entirely upon the official *volatile oil,* which is described below.

2. Oleum Carui.

Oil of Caraway.

Varieties.—The amount of the oil will vary according to the variety of caraway fruit from which it has been obtained, from about 4 to 7 per cent. The oil distilled in this country from home-grown fruits is regarded as the best.

General Characters and Composition.—Oil of caraway is colourless or pale yellow when recent, but by keeping it gradually becomes darker; it has the odour of the fruit, and a spicy somewhat acrid taste. It is dextrogyrate; its specific gravity is said to vary from about 0·916 to 0·946. It is composed of a hydrocarbon termed *carvene,* and an oxygenated oil, *carvol.*

Medicinal Properties.—Both the fruit and oil possess aromatic, stimulant, and carminative properties. They are principally employed as adjuncts to other medicines as corrective or flavouring agents.

Official Preparations.

1. Of CARUI FRUCTUS :—

Aqua Carui.	Pulvis Opii Compositus.
Confectio Piperis.	Tinctura Cardamomi Composita.
Oleum Carui.	Tinctura Sennæ.

Pulvis Opii Compositus is also used in the preparation of Confectio Opii ; and Tinctura Cardamomi Compositus and Tinctura Sennæ in several official preparations.

2. Of OLEUM CARUI :—

Confectio Scammonii.	Pilula Aloes Barbadensis.

3. PIMPINELLA ANISUM, *Linn.*

Anise.

(Bentley and Trimen's ' Medicinal Plants,' vol. ii. plate 122.)

Habitat.—Anise is a native of Egypt, Crete, Cyprus, and many islands of the Greek Archipelago. It is now also cultivated in several parts of Europe where the summer is warm enough for ripening its fruits; and in South America and India.

Official Parts or Products and Names.—1. ANISI FRUCTUS :—the dried fruit. 2. OLEUM ANISI:—the oil distilled in Europe from anise fruit (Pimpinella Anisum, *Linn.*); or in China from star-anise fruit (Illicium anisatum, *Linn.*).

1. Anisi Fructus.

Anise Fruit.

Varieties and Commerce.—Anise fruits are chiefly imported into this country from Spain, Germany, Russia, and Chili; the more ordinary varieties being the Alicante from Spain (which is the most esteemed), German, and Russian.

FIG. 27.—Anisi Fructus, *Anise Fruit.* A. Entire fruit. *a, a.* Primary ridges. *b, b.* Channels. *g.* Epigynous disk. *h.* Pedicel. B. Transverse section of the same. *a.* Primary ridges. *b.* Channels. *c.* Vittæ. *d, d.* Albumen. (Both figures enlarged.)

General Characters. — Anise fruits, excluding the Russian variety which is shorter, average about one-fifth of an inch in length. They are ovoid-oblong in form (*fig.* 27, A), greyish-brown, their whole surface is covered with short hairs, and their constituent mericarps are united and attached to a common stalk, the whole being capped with a small stylopod and two styles. Each mericarp has a flat face (*fig.* 27, B), and is marked on its dorsal surface by five pale slender entire ridges; and when divided transversely, the section exhibits fifteen or more small vittæ, which are principally evident in the channels (*fig.* 27, B). The fruit has an agreeable aromatic odour, and a sweetish spicy taste.

Principal Constituents.—The only important constituent is the official oil, which is described below.

2. Oleum Anisi.

Oil of Anise.

The amount of volatile oil obtainable from the different varieties has been estimated at from about 2 to 3 per cent. Its characters have been described under Illicium anisatum (page 23).

Medicinal Properties.—Both anise fruit and oil of anise have aromatic, carminative, and stimulant properties.

Official Preparations.

1. Of ANISI FRUCTUS :—
     Aqua Anisi. | Oleum Anisi.
2. Of OLEUM ANISI :—
     Essentia Anisi. *Dose.*—10 to 20 minims.
     Tinctura Camphoræ Composita. *Dose.* —· 15 minims to 1 fluid drachm.
     Tinctura Opii Ammoniata. *Dose.*—$\frac{1}{2}$ to 1 fluid drachm.

---

4. FŒNICULUM CAPILLACEUM, *Gilib.*

Fennel.   Sweet Fennel.

Synonym.—FŒNICULUM VULGARE, *Gaert.*

(Bentley and Trimen's 'Medicinal Plants,' vol. ii. plate 123.)

Habitat.—It is wild in most parts of Europe, except the north and east, and in Asia Minor, Persia, and India.   It is apparently wild in England.

Official Part and Name.—FŒNICULI FRUCTUS :—the dried fruit of cultivated plants.

Fœniculi Fructus.

Fennel Fruit.

Varieties and Commerce.—There are two principal varieties, which are known as *Roman* or *Sweet Fennel*, and *Saxon* or *German Fennel.* The former is obtained from plants cultivated at Nîmes in the South of France; and the latter more especially from plants cultivated near Weissenfels in Saxony. The first variety is the most esteemed.

General Characters.—The fruits are from about one-fifth to two-fifths of an inch in length. They are oblong or ovoid-oblong in form, (*fig.* 28, *a*), more or less curved or arched, smooth, greenish - brown or brown, and crowned by an evident stylopod and two styles. They have an aromatic odour, and a

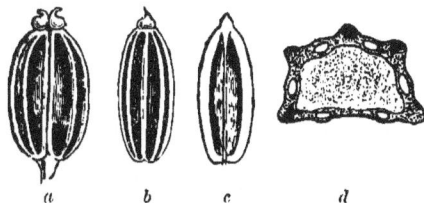

FIG. 28.— Fœniculi Fructus, *Fennel Fruit*. *a*. Entire fruit. *b*. Mericarp (dorsal surface). *c*. Mericarp (face or commissural surface). *d*. Transverse section of a mericarp. (All enlarged.)

more or less sweet aromatic taste. The entire fruit is commonly seen in commerce, but it is readily separable into its two mericarps (*fig.* 28, *b*, *c*), each of which has five prominent ridges (*fig.* 28, *b*), of which the lateral are the broadest, and four vittæ in the spaces between the ridges, and commonly two on the face (*fig.* 28, *d*). The Roman or Sweet Fennel is distinguished by its greater length, its more oblong form, and sweeter taste.

Principal Constituents. — Fennel fruits contain *sugar*, *fixed oil*, and a *volatile oil* which is the essential constituent. It is contained in a proportion varying from 3 to 4 per cent. It is composed of a hydrocarbon isomeric with oil of turpentine, and of the oxygenated oil, termed *anethol* or *anise-camphor,* which has been already mentioned as the principal constituent of oil of anise (page 23). It is more or less dextrogyre, according to the variety of fennel from whence obtained. Oil of sweet fennel is most valued. It has a decided sweet taste.

Medicinal Properties. — Like caraway, anise, and dill fruits, fennel fruits possess aromatic, carminative, and stimulant properties.

Official Preparations.

Aqua Fœniculi. | Pulvis Glycyrrhizæ Compositus.

---

5. PEUCEDANUM GRAVEOLENS, *Hiern.*

Dill.

Synonym.—ANETHUM GRAVEOLENS, *Linn.*

(Bentley and Trimen's ' Medicinal Plants,' vol. ii. plate 132.)

Habitat.—It is a native of the Mediterranean region, Southern Russia, and the Caucasian provinces. It also occurs more rarely as a corn-field weed in Northern Europe.

Official Parts or Products and Names.—ANETHI FRUCTUS :—the dried fruit. 2. OLEUM ANETHI :—the oil distilled in Britain from dill fruit.

1. Anethi Fructus.

Dill Fruit.

Varieties and Commerce. — Dill fruit of commerce is commonly imported from Central and Southern Europe. Indian dill fruit, which is distinguished by its mericarps being more convex, less winged, and narrower than the European fruits, is not included in the pharmacopœia description.

General Characters. — Dill fruits average about one-sixth of an inch long. They are broadly oval in outline, much compressed dorsally, rounded at both ends, smooth, crowned by a small stylopod, and surrounded by a broad membranous border. They have a brown colour, the membranous border being paler. In commercial fruits the mericarps are usually distinct (*fig.* 29, *a, b*), each being thin, flat, with three evident dorsal ridges (*fig.* 29, *a*), and two lateral ones, which expand into a thin wing (*fig.* 29, *c*) surrounding the mericarp. There are four broad vittæ in the grooves at the back of each mericarp, and two on the face. The odour and taste of dill fruit are agreeably aromatic, somewhat resembling caraway fruit.

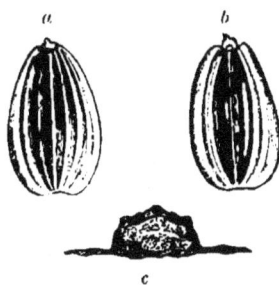

FIG. 29.—Anethi Fructus, *Dill Fruit. a.* Dorsal surface of mericarp. *b.* Mericarp (face or commissural surface). *c.* Transverse section of a mericarp. (All enlarged.)

Principal Constituents.—The only important constituent is the official oil, which is described below.

2. Oleum Anethi.

Oil of Dill.

General Characters and Composition.—Oil of dill is obtained by distillation in the proportion of from 3 to 4 per cent. It has a pale yellow colour, a pungent odour, and hot sweetish taste. It is dextrogyre, and has a specific gravity of about 0·87.

It is essentially composed of a hydrocarbon called *anethene,* also another hydrocarbon in small proportion, and an oxygenated oil which is identical with the *carvol* of oil of caraway.

Medicinal Properties.—Like the other aromatic official umbelliferous fruits and volatile oils, both dill fruit and oil of dill possess stimulant, aromatic, and carminative properties.

Official Preparations.

Of ANETHI FRUCTUS :—

Aqua Anethi. | Oleum Anethi.

---

6. CORIANDRUM SATIVUM, *Linn.*
Coriander.

(Bentley and Trimen's ' Medicinal Plants,' vol. ii. plate 133.)

Habitat.—Scarcely known in a wild state, but apparently indigenous to the Mediterranean and Caucasian regions. In this country and in other parts of Europe it is only found as an escape from cultivation.

Official Parts or Products and Names.—1. CORIANDRI FRUCTUS :—the dried ripe fruit. 2. OLEUM CORIANDRI :— the oil distilled in Britain from coriander fruit.

1. Coriandri Fructus.
Coriander Fruit.

Cultivation and Commerce. — Coriander fruits are obtained to a small extent from plants cultivated in Britain, more especially in Essex. Coriander is also cultivated in various parts of Continental Europe, and in Northern Africa and India. The characters given in the pharmacopœia apply more particularly to the European kinds. Indian coriander fruits are larger and somewhat elongated in form.

General Characters.—Nearly globular in form (*fig.* 30, *a*), averaging about one-fifth of an inch in diameter, although sometimes less ; the two constituent mericarps, which are concave on their face (*fig.* 30, *c*), closely united, and enclosing a lenticular cavity (*fig.* 30, *c*, *d*) ; the whole crowned with the calyx-teeth and stylopod, and in some instances by the two

slender spreading styles (*fig.* 30, *a*).　The fruit is somewhat hard, and of a brownish-yellow colour.　Each mericarp is faintly ribbed longitudinally with five very sinuous primary and four secondary ridges (*fig.* 30, *b*), the latter being the more pro-
minent, but only very slightly raised.　There are no vittæ on the dorsal surface, but each mericarp has two brown curved vittæ on its face or commissural surface (*fig.* 30, *d*).　Coriander fruit has an agreeable mild aromatic taste, and when bruised a pleasant odour.

Fig. 30.—Coriandri Fructus, *Coriander Fruit.* *a.* Entire fruit. *b.* Mericarp (dorsal surface). *c.* Commissural surface of a mericarp. *d.* Transverse section of the fruit. (All enlarged.)

Principal Constituents.—The essential constituent is the official *volatile oil* described below.　The fruit is also said to contain about 13 per cent. of *fixed oil.*

2. Oleum Coriandri.

Oil of Coriander.

The oil is directed to be distilled in Britain ; and before the fruit is submitted to distillation, it should be crushed or bruised.　The amount obtainable varies from about $\frac{1}{2}$ to 1 per cent.

General Characters and Composition.—Oil of coriander is colourless or pale yellow, with the odour of the fruit, and a mild aromatic taste.　It is dextrogyre, and has a specific gravity of about 0·86 to 0·87.　Oil of coriander has a com-position isomeric with *borneol.*

N

Medicinal Properties.—Both coriander fruit and oil of coriander possess aromatic, stimulant, and carminative properties, like caraway and the other aromatic umbelliferous fruits, and may therefore be used in similar cases.

Official Preparations.

1. Of CORIANDRI FRUCTUS :—

Confectio Sennæ.	Tinctura Sennæ, which
Oleum Coriandri.	is also used in the
Syrupus Rhei.	preparation of Mistura
Tinctura Rhei.	Sennæ Composita.

2. Of OLEUM CORIANDRI :—Syrupus Sennæ.

---

7. CUMINUM CYMINUM, *Linn.*

Cummin or Cumin.

(Bentley and Trimen's 'Medicinal Plants,' vol. ii. plate 134.)

Habitat.—Scarcely known in a wild state ; but probably originally indigenous in North-Eastern Africa and Western Asia. It has been cultivated from the earliest times in the countries bordering the Mediterranean, and in India.

Part Used and Name.—CUMINI FRUCTUS :—the dried ripe fruit.

(*Not Official.*)

Cumini Fructus.

Cumin Fruit.

Commerce.—Cumin is generally obtained, in Britain, from Mogador, Malta, and Sicily ; but it is also largely produced in India.

General Characters.—Cumin fruits are oval-oblong in form, averaging about one-quarter of an inch in length, somewhat tapering to each end, slightly compressed laterally, brown, more or less covered with rough hairs (or very rarely the hairs are entirely absent), and capped by the persistent calyx-teeth. In taste and odour they somewhat resemble caraway fruits, but are far less agreeable. The two constituent mericarps are not readily separable, and each is marked by five primary and four secondary ridges, the latter being the more prominent; and between each primary ridge, beneath each secondary ridge, there is a small vitta, and two vittæ occur on the face.

Principal Constituents.—The properties are due to a *volatile oil*, of which they yield from about 3 to 4 per cent. Oil of cumin has a pale yellow colour, a strong acrid aromatic taste, and disagreeable odour. It is dextrogyre, and has a specific gravity of about 0·92. It is essentially composed of the hydrocarbon *cymol* or *cymene*, and an oxygenated oil named *cuminol* or *cuminaldehyd*.

Medicinal Properties.—Aromatic, carminative, and stimulant, like caraway; but it is rarely or ever used medicinally in this country, caraway and other aromatic umbelliferous fruits being more agreeable. It is chiefly employed in veterinary medicine, and as a condiment.

---

8. FERULA SUMBUL, *Hook. fil.*
Sumbul.

Synonym.—EURYANGIUM SUMBUL, *Kauffmann.*

(Bentley and Trimen's ' Medicinal Plants,' vol. ii. plate 129.)

Habitat.—This plant was originally discovered on the mountains to the south-east of Samarkand between Russian Turkestan and Bucharia.

Official Part and Name.—SUMBUL RADIX :—the dried transverse sections of the root.

Sumbul Radix.
Sumbul Root.

Commerce.—Sumbul root is the produce of Bucharia and Turkestan. It is imported into Europe and elsewhere by way of Russia.

General Characters.—The slices vary very much in size,

but are usually from about one to three inches in diameter, and from three-quarters of an inch to one inch or more in thickness; but in some cases pieces may be found as much as five inches in diameter, and in others not thicker than a common quill. Externally the slices are covered by a dusky-brown transversely wrinkled papery bark; and in the case of those derived from the crown of the root, they are beset with short bristly fibres. Internally they are dirty yellowish-brown, mottled with whitish patches; and when the surface is examined with a lens it exhibits, more particularly towards the circumference, spots of exuded resin. Their structure is spongy, coarsely fibrous, dry, and farinaceous-looking. The odour is strong, pleasant, and musk-like, which is long retained; hence the name of *musk-root* by which it is commonly known. Their taste is bitter and aromatic.

Substitution.—In the British Pharmacopœia of 1867, sumbul root was stated to be derived from Russia and India; the latter being distinguished as *Indian Sumbul* or *Bombay Sumbul,* but this is now known to consist of dried prepared slices of ammoniacum root, from *Dorema Ammoniacum.* They are readily known from true sumbul root by their closer texture, greater firmness, yellowish-red colour, and less marked musky odour. This spurious sumbul root has of late years also appeared in commerce.

Principal Constituents.—Sumbul root yields about 9 per cent. of a soft *resin,* which has a musky odour, and is soluble in ether; it is that constituent in which the activity of sumbul root appears essentially to reside. The other constituents are about $\frac{1}{3}$ per cent. of a bluish *volatile oil*; $\frac{3}{4}$ per cent. of *angelic acid* (*sumbulic acid*), which is crystallisable, and is associated with a trace of *valerianic acid, starch,* etc. *Umbelliferone* may be also obtained by dry distillation from sumbul resin.

Medicinal Properties.—Nervine stimulant like valerian, and antispasmodic.

Official Preparation.—Tinctura Sumbul. *Dose.*— 10 to 30 minims.

9. FERULA NARTHEX, *Boiss.*
Thibetan Asafœtida.

Synonym.—NARTHEX ASAFŒTIDA, *Falconer.*

(Bentley and Trimen's ' Medicinal Plants,' vol. ii. plate 126.)

Habitat.—This plant was discovered by Falconer in 1838, in the valley of Astore or Hussorah in Western Thibet ; and at present we have no positive evidence of its having elsewhere been found.

10. FERULA SCORODOSMA, *Benth. & Hook. fil.*
Persian Asafœtida.

Synonyms.—SCORODOSMA FŒTIDUM, *Bunge* ; FERULA FŒTIDA, *Regel.*

(Bentley and Trimen's ' Medicinal Plants,' vol. ii. plate 127.)

Habitat.—Afghanistan, Turkestan, and Eastern Persia.

Official Product and Name.—ASAFŒTIDA :—a gum-resin obtained by incision from the living root of the above two plants ; and probably other species.

Asafœtida.
Asafœtida.

Collection, Source, and Commerce.—The mode of collection has recently been described by Dr. Aitchison as follows :—
' The men begin their work by laying bare the root-stock to a depth of a couple of inches of those plants only which have not as yet reached their flower-bearing stage. They then cut off a slice from the top of the root-stock, from which at once a quantity of milky juice exudes, which my informant told me was not collected then. They next proceeded to cover over the root by means of a domed structure, of from six to eight inches in height, called a

khora, formed of twigs and covered with clay, leaving an opening towards the north, thus protecting the exposed root from the rays of the sun. The drug collectors return in about five or six weeks' time, and it was at this stage that the process of collecting came under my personal observation. A thick gummy, not milky, reddish substance now appeared in more or less irregular lumps upon the exposed surface of the root, which looked to me exactly like the ordinary asafœtida of commerce, as employed in medicine. This was scraped off with a piece of iron hoop, or removed along with a slice of the root, and at once placed in a leather bag, the tanned skin of a kid or goat. My guide informed me that occasionally the plant was operated upon in this manner more than once in the season. The asafœtida was then conveyed to Herat, where it usually underwent the process of adulteration with a red clay, *táwah*, and where it was sold to certain export traders, called *Kákri-log*, who convey it to India.'

Commercial asafœtida is sometimes stated to be the produce entirely of Afghanistan, and as the recent investigations of Aitchison have shown that this is derived from *Ferula Scorodosma*, and not from *F. Narthex*, as formerly supposed, we have now no positive evidence in reference to the supply of commercial asafœtida from the latter plant. Dymock, however, says that some of the asafœtida of European commerce is also produced in the province of Laristan in Persia, and this also appears to be the produce of *Ferula Scorodosma*. Whatever may be the source of Asafœtida, it subsequently reaches Bombay, from whence the drug is exported to Europe and other parts of the world.

Varieties.—Commercial asafœtida is commonly seen in mass, rarely in tears, and sometimes as a fluid honey-like substance, which by keeping becomes brown. But besides these, other kinds are known in India ; they are not, however, found in European trade.

General Characters and Tests. Usually in masses of

irregular forms, varying in consistence and size, and composed of whitish tears agglutinated together by dark yellowish-grey or brownish softer material. When broken or cut, the exposed surface has usually a more or less amygdaloid appearance (*amygdaloid asafœtida*), the fractured tears being opaque and milk-white or yellowish at first, but changing gradually by exposure to air to purplish-pink or reddish-pink, and finally to dull yellowish-brown. But in some specimens, in consequence of the deficiency of tears, and the larger proportion of connecting material, the amygdaloid character is not very evident. Asafœtida readily softens by heat and then becomes sticky, but in cold weather it is so hard and brittle that it may be powdered. It has a bitter acrid alliaceous taste ; and a powerful and persistent alliaceous odour. When triturated with water it forms a whitish emulsion. The freshly fractured surface of a tear when touched with nitric acid assumes for a short time a fine green colour. It should yield not more than 10 per cent. of ash; and from 50 to 60 per cent. should be soluble in rectified spirit.

Principal Constituents.—Asafœtida is essentially composed of *volatile oil*, *resin*, and *gum*, in varying proportions in different specimens. The resin and volatile oil are the active constituents ; the former generally forming more than one-half, and the latter from 4 to 9 per cent. of the drug. The volatile oil is dextrogyre, has a yellow colour, very alliaceous odour from containing a large amount (20 to 25 per cent.) of sulphur, and its specific gravity at about 76° Fahr. is 0·950. On dry distillation asafœtida also yields a little *umbelliferone.*

Medicinal Properties.—Stimulant, powerfully antispasmodic, carminative, and expectorant. It is also sometimes regarded as emmenagogue and anthelmintic.

Official Preparations.

Enema Asafœtidæ.	Pilula Asafœtidæ Composita.
Pilula Aloes et Asafœtidæ.	Spiritus Ammoniæ Fœtidus.
Tinctura Asafœtidæ.	

11. FERULA GALBANIFLUA, *Boiss.* & *Buhse.*
Galbanum.

(Bentley and Trimen's ' Medicinal Plants,' vol. iii. plate 12S.)

Habitat.—Afghanistan and Northern Persia.

12. FERULA RUBRICAULIS, *Boiss.*

Habitat.—Persia and Afghanistan.

Official Product and Name.—GALBANUM :—a gum-resin obtained from the above two plants; and probably other species.

Galbanum.
Galbanum.

Collection and Commerce.—The mode of obtaining this drug is not in all cases accurately known, for while by some it is stated to be derived from the root or lower part of the stem by incision; others, as Buhse and Aitchison, who have personally visited some of the districts in which it is collected, state that the juice exudes spontaneously, more especially from the lower part of the stem, and at the points of insertion of the leaves. Galbanum is chiefly brought into commerce from Eastern Europe, or by way of India; but considerable quantities appear also to reach Russia by Astrachan and Orenburg.

General Characters.—Galbanum occurs in tears (*galbanum in tear*), or in masses of agglutinated tears (*galbanum in mass*). The tears are roundish or irregular in form, and usually vary in size from that of a lentil to a hazel-nut, although rarely exceeding that of a pea ; but in some cases larger tears may be found. They are light yellowish-brown, orange-brown, or yellowish-green in colour externally ; and internally yellowish or milk-white. They are more or less translucent ; usually rough and dirty on the surface ; hard and brittle in cold weather, but softening in the summer, and by the heat of the hand becoming ductile and sticky. The masses, which frequently contain pieces of root, stem, and other impurities, vary in consistence and size; but are usually compact and hard, irregular in form, and yellowish-brown, dark brownish-yellow, or rarely greenish in colour. Galbanum has a peculiar, somewhat aromatic, and not disagreeable odour; and a bitter, unpleasant, feebly acrid, and somewhat alliaceous taste. When triturated with water it forms a whitish emulsion. At least 60 per cent. should be soluble in rectified spirit. In rare cases galbanum is found in a soft, almost fluid state.

Principal Constituents.—Galbanum is a *gum-resin*, and yields on an average from 60 to 65 per cent. of *resin*; 6 to 9 per cent. of *volatile oil*; and from 20 to 30 per cent. of *gum*. Its properties are due to the resin and volatile oil. It also yields about 0·8 per cent. of *umbelliferone*, a substance likewise obtainable, as we have seen, from sumbul resin and asafœtida, but in smaller proportion. If a small fragment of galbanum be immersed for an hour or two in hot water, on the addition of a drop of ammonia to the infusion a blue fluorescence is immediately produced, which disappears again on the addition of an acid. This reaction is due to *umbelliferone*; and hence does not take place with ammoniacum, in consequence of its not being present in that drug. The *volatile oil* is dextrogyrate like that of asafœtida, but it does not contain sulphur like it.

Medicinal Properties. — Stimulant, expectorant, and slightly antispasmodic, when given internally; and when used externally, as in galbanum plaster, it acts as a mild stimulant to promote the resolution of tumours, etc.

Official Preparations.

Emplastrum Galbani. | Pilula Asafœtidæ Composita.

13. DOREMA AMMONIACUM, *Don.*

Ammoniacum.

(Bentley and Trimen's ' Medicinal Plants,' vol. ii. plate 131.)

Habitat.—Afghanistan and Persia.

Official Product and Name.—AMMONIACUM :—a gum-resinous exudation from the stem after being punctured by beetles.

Ammoniacum.

Ammoniacum.

Collection and Commerce.—Commercial ammoniacum, although sometimes stated to be obtained by incising the stem, appears to be an exudation entirely caused by the puncture of beetles. As the juice exudes and hardens, the ammoniacum partly adheres to the stem, and partly falls on the ground ; and after collection, it is forwarded from Afghanistan and Persia, where it is obtained for commercial purposes, to Bombay, and thence to other parts of the world.

Characters and Tests.—In commerce, ammoniacum occurs in separate tears (*ammoniacum in tear*), or in masses of more or less agglutinated tears (*lump ammoniacum*).

Ammoniacum in tear, which is the purest form of ammoniacum, consists of the separate tears, which are roundish or somewhat irregular in form, and vary in size usually from that of a coriander fruit to a cherry; but sometimes larger. Externally, when recent, they have a pale yellowish-brown colour, but darkening by long keeping to cinnamon-brown ; internally, they are milky-white and opaque. At ordinary temperatures they are hard and brittle, and break with a dull waxy fracture, but they are readily softened with heat. Ammoniacum has a faint peculiar non-alliaceous odour, and a bitter acrid taste.

When triturated with water it forms a nearly white emulsion. It is coloured yellow by caustic potash ; and a solution of chlorinated soda gives it a bright orange hue.

Lump ammoniacum is in nodular masses of varying sizes and forms, and should consist of tears agglutinated together, so that when broken they present an amygdaloid appearance. The characters generally are, therefore, the same as ammoniacum in tear ; but the masses are commonly less pure than the separate tears from being more or less mixed with the mericarps of the fruit and other extraneous substances.

Principal Constituents.—Ammoniacum, like asafœtida and galbanum, is a mixture of *gum, resin,* and *volatile oil* ; the properties being due to the two latter constituents. The resin is said to form 70 per cent. of the drug; and the oil, according to Vigier, about 1·8 per cent., but the authors of ' Pharmacographia' found only $\frac{1}{3}$ per cent. The oil is free from sulphur, and has the odour of ammoniacum. Unlike the other official umbelliferous gum-resins, it yields no *umbelliferone.*

Medicinal Properties.—Stimulant and expectorant when given internally ; it is also employed externally as a local irritant.

Official Preparations.

Emplastrum Ammoniaci cum Hydrargyro.	Mistura Ammoniaci.
	Pilula Scillæ Composita.
Emplastrum Galbani.	Pilula Ipecacuanhæ cum Scilla.

SUB-CLASS II.—GAMOPETALÆ OR COROLLIFLORÆ.

SERIES I.—*INFERÆ OR EPIGYNÆ.*

ORDER I.—CAPRIFOLIACEÆ.

SAMBUCUS NIGRA, *Linn.*

Elder.   Bore Tree.

(Bentley and Trimen's ' Medicinal Plants,' vol. ii. plate 137.)

Habitat.—All parts of Europe, except in the north-east, and extending into North Africa, the Caucasian provinces, and Southern Siberia. It is regarded as truly wild in England and Ireland, but not really indigenous to Scotland.

Official Part and Name.—SAMBUCI FLORES :—the fresh flowers.

Sambuci Flores.

Elder Flowers.

Collection and Preservation.—The flowers are directed to be obtained from indigenous plants ; and for use in pharmacy the fresh flowers are to be separated from the stalks, or an equivalent quantity of the flowers preserved while fresh with common salt. The latter are commonly known as *pickled flowers.*

General Characters.—In corymbose or flat-topped cymes, from about four to seven inches across, the main stalk dividing at first into from three to five branches (usually five), which subdivide once or more by threes or fives, and then are frequently and irregularly dichotomous ; all perfectly smooth and without bracts. Flowers small ; calyx superior, four- or five-toothed, usually five ; corolla about one-fifth of an inch wide, flat, rotate, five-sected, segments rounded, creamy-white ; stamens five, inserted into the short tube of the corolla, and alternate with the segments ; ovary projecting above and crowned by three sessile stigmas.   Odour fragrant, but somewhat sickly ; taste bitterish.

Principal Constituents.—The essential constituent is a *volatile oil*, of which elder flowers yield when distilled with water a very small quantity.   This oil is colourless, of a butter-like consistence, and with the strong odour of the fresh flowers.   Its specific gravity is less than that of water, and it is readily decomposed by exposure to the air.

Medicinal Properties.—Elder flowers are somewhat stimulant and sudorific, when employed for internal administration.   But in this country they are only used in the form of elder flower ointment and of the official elder flower water, as a cooling application to inflamed surfaces ; and as a menstruum in lotions.

Official Preparation.—Aqua Sambuci.

ORDER 2.—RUBIACEÆ.

1. UNCARIA GAMBIER, *Roxb.*

Gambier.

(Bentley and Trimen's ' Medicinal Plants,' vol. ii. plate 139.)

Habitat.—It is a native of the countries bordering on the Straits of Malacca, more especially of the islands at their eastern end; and also of Ceylon.

Official Product and Name.—CATECHU :—an extract of the leaves and young shoots.—Uncaria acida, *Roxb.*, which is probably only a variety of U. Gambier, is also used in the preparation of catechu.

Catechu.

Synonym.—Catechu Pallidum.

Commercial Kinds of Catechu.—Besides this kind of catechu, which is *alone official* in the British Pharmacopœia, other kinds are known in commerce, more especially that derived from *Acacia Catechu,* which has already been described under the name of *Catechu Nigrum.* The latter name at once distinguishes it from the official *Catechu Pallidum.*

Cultivation, Preparation, and Commerce.—The plants yielding the official catechu are largely cultivated at Singapore, and on the island of Bintang and other islands of the Rhio-Lingga Archipelago lying south-east of Singapore. It is obtained by boiling the leaves and young shoots in water, and then evaporating the decoction to a proper consistence ; after which the thickened mass is placed in shallow square boxes, and when sufficiently hardened is cut into cubes and dried in the shade. It is chiefly exported from Singapore.

Characters and Tests.—*Catechu, Pale Catechu, Gambier,* or *Terra Japonica,* under all of which names it is commonly known, occurs in separate cubes, or in masses of variable sizes formed of more or less agglutinated cubes. The separate cubes, which are frequently chipped or broken, are usually about an inch square on each side, deep reddish-brown externally, pale cinnamon-brown internally, dry, friable, and when fractured the surface presents a dull earthy appearance, and hence the name *Terra Japonica* by which it was originally known, under the impression that it was an earthy substance obtained from Japan. It has no odour, but a bitterish and very astringent taste at first, and subsequently sweetish. Examined by the microscope it presents myriads of very minute acicular crystals of *catechin* or *catechuic acid,* of which it mainly consists. It is entirely soluble in boiling water ; and the decoction when cold is not rendered blue by iodine, indicating the absence of starch.

Principal Constituents.—Its principal constituent is *catechin* or *catechuic acid,* which, as noticed above, may be seen under the microscope in the form of very small acicular crystals. It also contains a small quantity of *catechu-tannic acid* ; and a yellow colouring principle named *quercetin.* Its composition is therefore essentially the same as that of cutch or black catechu from *Acacia Catechu* (*see* Catechu Nigrum); but it contains much more catechuic acid than it.

Medicinal Properties.—It is a valuable astringent, and largely employed both internally and externally as a medicinal agent. It is more soluble than black catechu, and more powerful than kino.

Official Preparations.

Infusum Catechu. | Tinctura Catechu.
Pulvis Catechu Composita. | Trochisci Catechu.

2. CINCHONA, *Linn.*

The following species of Cinchona are specially referred to by name in the British Pharmacopœia, namely, the more important official species—*Cinchona succirubra*, Pavon ; *Cinchona Calisaya*, Weddell ; *Cinchona officinalis*, Linn. ; and *Cinchona lancifolia*, Mutis : all of which are figured in Bentley and Trimen's 'Medicinal Plants,' vol. ii. plates 140, 141, 142, and 143.

The official parts and products of Cinchona will be described under their two pharmacopœia headings, of 'Cinchonæ Rubræ Cortex,' and 'Cinchonæ Cortex.'

1. Cinchonæ Rubræ Cortex.

Red Cinchona Bark.

This is officially described as 'the dried bark of the stem and branches of cultivated plants of Cinchona succirubra, *Pavon.*'

Habitat, Cultivation, and Preparation.—The plant from which this bark is obtained was formerly a native of the province of Huaranda, in Ecuador, but is now only found growing wild on the western slopes of Chimborazo, near Guayaquil It is, however, cultivated to an enormous extent in the Neilgherry Hills on the south-west coast of India and in the Madras Presidency, in the valleys of the

Himalayas in British Sikkim, and in Ceylon, Java, Jamaica, etc.; and this bark from cultivated plants is now *alone* official for the galenical preparations of Cinchona in the British Pharmacopœia. But the barks of other species of *Cinchona*, as well as it, and of some species of *Remijia*, DC., may also be employed for obtaining the official salts of the Cinchona alkaloids (*see* Cinchonæ Cortex).

The quills and incurved pieces of this bark, described in the pharmacopœia, are obtained by making vertical and transverse incisions through the bark, according to the length and breadth of the pieces required; these are afterwards separated from the wood beneath with a common knife or other suitable instrument, by introducing the point of the knife as closely as possible to the surface of the wood, and then peeling off the bark. This bark is subsequently dried, when it curls up into hollow cylinders or quills, or in more or less incurved pieces.

A large proportion of the bark is, however, now imported in the form of shavings or strips loosely aggregated together, or pressed into more or less compact masses. These shavings are not, however, suitable for general use by the pharmacist, as they are difficult to recognise and assign to the particular species from whence they have been derived. These strips are obtained in certain districts by the process of cultivation adopted. Thus, it is found that by removing longitudinal strips of the bark from the stem or branches in the same manner as cork is obtained in the South of Europe, and carefully covering the exposed surface with moss or other suitable material, the bark is renewed, and is at least equally rich in alkaloids; and thus by taking successive strips of bark in different years, the old bark becomes renewed by younger bark, and the trees made to yield continuously a supply of rich bark. This process of mossing, as it is termed, cannot however be adopted in all districts, on account of the insects, chiefly ants, getting beneath the moss, and destroying the young bark as rapidly as it is formed.

General Characters.—In quills or more or less incurved pieces, coated with the periderm, and varying in length from usually a few inches to a foot or more—the bark itself being from about one-tenth to a quarter of an inch thick, or rarely more. The outer surface is more or less rough from longitudinal furrows and ridges, or transverse cracks, annular fissures, and warts, and brownish or reddish-brown in colour; and the inner surface brick-red or deep reddish-brown, and irregularly and coarsely striated. The fracture is nearly close in the smaller quills, but finely fibrous in the larger ones; and the powder brownish or reddish-brown. It has no marked odour; but the taste is bitter and somewhat astringent.

Test.—When used for purposes other than that of obtaining the alkaloids or their salts, it should yield between 5 and 6 per cent. of total alkaloids, of which not less than half shall consist of quinine and cinchonidine, as estimated by the methods given in the pharmacopœia.

2. Cinchonæ Cortex.
Cinchona Bark.

This is officially described as 'the dried bark of Cinchona Calisaya, *Weddell*; Cinchona officinalis, *Linn.*; Cinchona succirubra, *Pavon*; Cinchona lancifolia, *Mutis*; and other species of Cinchona from which the peculiar alkaloids of the bark may be obtained.' It is also stated that 'salts of quinine and cinchonine may be obtained from some species of Remijia, *DC.*'

1. CINCHONA CALISAYA.—This species was mentioned in the Pharmacopœia of 1867, as the botanical source of the official Yellow Cinchona Bark of that volume, and was then principally directed to be used in the preparations; but at present the bark obtained from it is only referred to as a source of the Cinchona alkaloids.

Yellow Cinchona Bark or *Calisaya Bark* is found in flat pieces or in quills. *Flat Calisaya* is in flattish (*fig.* 31) pieces of varying length and breadth, sometimes a foot or

o

more, and three or four inches in width ; and in thickness it varies from about one-fifth to two-fifths of an inch. The outer layers are almost entirely absent, so that the pieces consist essentially of liber, which is of a compact texture, and externally of a tawny-yellow or rusty orange-brown colour, frequently interspersed with darker patches. Externally, it is also marked with shallow longitudinal depressions, which are termed *digital furrows*, separated by somewhat sharp projections. The inner surface has a wavy and closely and finely striated appearance. Its transverse fracture is finely fibrous (*fig.* 31), the fibres being very short, stiff, and readily detached, so that when this bark is handled it causes much irritation. It has a feeble somewhat aromatic odour, and a well-marked bitter taste. Its powder is of a light cinnamon-brown colour.

The *quills* are either single or rolled up at both margins so as to form double quills, and of varying length. In diameter they vary from about half an inch to nearly two inches ; the bark itself being from one-sixteenth to one-eighth of an inch or more thick. They are always covered with a periderm of a brownish colour naturally, but are

Fig. 31. - - Flat Calisaya Bark, showing its very finely fibrous fracture, and digital furrows.

frequently greyish or silvery-white from attached lichens ; the periderm is more or less rough on its outer surface from being marked with deep longitudinal furrows and transverse cracks, the latter of which in the larger quills often form nearly complete rings around the quills, and have their edges somewhat raised and everted. The inner surface is dark cinnamon-brown and finely striated from the liber-fibres. Its fracture is finely fibrous like the flat bark, and its taste also similar to it.

This bark has been always regarded as the best of all the cinchona barks, and was formerly readily obtainable from Bolivia and South Peru ; but our supplies, which are irregular, are now essentially derived from plants under cultivation, and more especially in Java and India. Recently, however, some fine bark has been again received from Bolivia. One variety of *Cinchona Calisaya*, which has been more particularly cultivated in Java, but is now being extensively tried in India, and known as var. *Ledgeriana*—but probably, as regarded by Dr. Trimen, a distinct species (*C. Ledgeriana*)—yields a bark of extraordinary richness in alkaloids, as much as 13·25 per cent. of quinine having been obtained from it.

2. CINCHONA OFFICINALIS.—This species yields the Pale Cinchona Bark which was official in the British Pharmacopœia of 1867, and its botanical source there referred to as *C. Condaminea*. It is now only mentioned as one of the sources of the official salts of the Cinchona alkaloids. *Pale Cinchona Bark* is also known as *Loxa Bark* or *Crown Bark*, and several varieties have been described by pharmacologists. They were originally obtained from Loxa in Ecuador, but at present our supplies of pale bark are essentially or entirely derived from plants cultivated in India, Ceylon, Java, etc.

Pale Bark is only found in quills coated with periderm. These quills are single or double, vary in length from commonly a few inches to a foot or more, and in diameter from one-eighth to three-quarters of an inch, while the bark itself, although sometimes not thicker than cartridge paper, is in other specimens one-tenth of an inch or even more. The outer surface has a dark-grey or blackish-brown colour, and is frequently spotted with silver-grey patches from the presence of lichens. The smaller quills are only slightly marked on their outer surface with longitudinal and transverse wrinkles or furrows ; but the larger ones are rough from the presence of deep transverse or annular cracks. The inner surface has a bright cinnamon-brown or reddish-brown colour. Pale bark has a somewhat fibrous fracture internally, particularly the larger quills ; and its powder is pale brown. The odour is slight and peculiar ; and the taste bitter and astringent.

3. CINCHONA LANCIFOLIA.—This was only mentioned in the Pharmacopœia of 1867 as yielding one of the barks from which sulphate of quinia (quinine) was directed to be obtained.

In commerce this bark is commonly known as *Coquetta, Coqueta, Carthagena,* and *Columbian Bark.* It is imported from New Granada, or from plants cultivated in India, Java, etc. Coquetta bark is found both in quills and flattish or incurved pieces. It is readily distinguished, when in quills, by the periderm being nearly smooth and of a brownish or yellow colour naturally, but frequently mottled with whitish lichens ; by its long fibrous or stringy fracture ; and its feebly bitter taste. The flattish and incurved pieces are with or without periderm ; in the former case the outer surface is nearly smooth, and in all cases the fracture is very stringy, and the taste but feebly bitter.

Principal Constituents of Cinchona Barks.—The most important constituents of cinchona barks are the alkaloids which exist, combined with *kinic acid,* in the form of *kinates.* The principal of these are *Quinine, Quinidine* or *Conquinine, Cinchonine, Cinchonidine,* and *Quinamine.* These alkaloids were till recently regarded as peculiar to cinchona barks; but quinine, quinidine, and cinchonine, together with two other allied alkaloids, known as *cupreine* and *homoquinine,* have now been found in *Cuprea barks.* These latter barks are derived from some species of *Remijia,* a genus closely allied to *Cinchona,* and formerly combined with it. The species of *Remijia* known to yield these alkaloids are *R. pedunculata* and *R. Purdieana,* but they probably exist in other species.

The following salts of three of the above alkaloids are official in the British Pharmacopœia: -- *Cinchonidinæ Sulphas,*

Cinchoninæ Sulphas, Quininæ Hydrochloras, and *Quininæ
Sulphas.* These may be obtained from the barks of the
species of Cinchona mentioned above under the head of
' Cinchonæ Cortex ;' and also, so far as the salts of quinine
and cinchonine are concerned, from some species of *Remijia.* Some other constituents of cinchona barks are
cincho-tannic acid and its derivative *cinchona-red,* and *kinovic
acid.*

Medicinal Properties.—The cinchona barks possess
powerful antiperiodic, tonic, and antipyretic properties; and
are also astringent. The first three properties are especially
due to their alkaloids, and the latter to cincho-tannic acid
and cinchona-red. The essential difference therefore between
the action of the salts of the alkaloids, and of the cinchona
barks, rests in the astringency of the latter.

Official Preparations.

1. Of CINCHONÆ RUBRÆ CORTEX :—

Decoctum Cinchonæ.	Mistura Ferri Aromatica.
Extractum Cinchonæ Liqui-dum.	Tinctura Cinchonæ.
Infusum Cinchonæ Acidum.	Tinctura Cinchonæ Com-posita.

2. Of CINCHONÆ CORTEX :—

Cinchonidinæ Sulphas. *Dose.*—1 to 10 grains.
Cinchoninæ Sulphas. *Dose.*—1 to 10 grains.
Quininæ Hydrochloras. *Dose.*—1 to 10 grains.
Quininæ Sulphas. *Dose.*—1 to 10 grains.

3. Of QUININÆ HYDROCHLORAS :—Tinctura Quininæ.

4. Of QUININÆ SULPHAS :—

Ferri et Quininæ Citras. | Tinctura Quininæ Ammoniata.
Vinum Quininæ.

3. COFFEA ARABICA, *Linn.*
Coffee.

(Bentley and Trimen's ' Medicinal Plants,' vol. ii. plate 144.)

Habitat.—Native of the tropical parts of Africa. It is also now cultivated in nearly all tropical and sub-tropical countries.

Official Product and Name.—CAFFEINA :—an alkaloid usually obtained from the dried leaves of Camellia Thea, *Link*; or the dried seeds of Coffea arabica, *Linn.* (See *Camellia Thea,* page 59.)

Caffeina.
Caffeine.

Synonyms.—Caffeia ; Theina ; Guaranina.

For a description of Caffeine reference must be made to the British Pharmacopœia, and to special works on Chemistry; but the importance of Coffee as an unfermented beverage renders it desirable that some notice should be given of its nature, preparation, and properties.

Nature and Preparation of Coffee Seeds. — Each fruit contains two seeds, which are enclosed in a parchment-like or papery endocarp, from which they are separated by drying, and by the action of peeling and winnowing mills. In commerce coffee is found in two states, *raw* and *roasted. Raw Coffee* consists of the seeds, which are erroneously termed coffee berries, separated from the endocarp, and also in part from their testa ; and *roasted coffee* consists of the same seeds which have been submitted to the process of roasting. In the latter process the testa separates from the seed, and is commonly termed by the roasters ' flights ' or ' the fibre;' and the remaining portion, which essentially consists of the albumen, is that which is used in the preparation of the un-fermented beverage known as coffee.

Principal Constituents.—The principal constituents of raw coffee are *caffeine*, which is identical with *theine* (see *Camellia Thea*), *caffeo-tannic acid*, a peculiar kind of *sugar*, and about 13 per cent. of *gluten.* The recent experiments of Paul and Cownley show that the amount of caffeine in samples of raw coffee drawn from various sources and dried at 212° F. is from 1·20 to 1·39 per cent. By the process of roasting it undergoes certain changes, the more important of which are said to be the formation of a *brown bitter principle,* and a *volatile oil* called *caffeone* ; but no change is produced in the alkaloid itself. As a result of these changes the seeds swell to nearly twice their original volume, although they lose from 15 to 20 per cent. of their weight ; and they acquire a fragrant odour, and a decidedly bitter taste. The effects of roasted coffee appear to be due to the combined influence of caffeine, caffeo-tannic acid, the bitter principle, and volatile oil.

Medicinal Properties.—In its effects and uses coffee closely resembles tea, but its astringent properties are much less (see *Tea*). Medicinally it is only used in the form of its official alkaloid caffeine, and of the citrate and other salts of caffeine, which are in some use as nervine stimulants.

Official Preparation of Caffeina.

Caffeinæ Citras.   *Dose.*—2 to 10 grains.

---

4. CEPHAËLIS IPECACUANHA, *A. Rich.*
Ipecacuanha.

(Bentley and Trimen's ' Medicinal Plants,' vol. ii. plate 145.)

Habitat.—Brazil, New Granada, and probably Bolivia.

Official Part and Name.—IPECACUANHA :—the dried root.

Ipecacuanha.
Ipecacuanha.

Collection, Preparation, and Commerce.—In Brazil, where Ipecacuanha is especially collected, the plant is known under the name of *Poaya,* and hence the collectors are called *Poayeros.* Ipecacuanha is collected more or less all the year round, and to this circumstance the difference in its colour externally is more especially due. A Poayero collects the roots by grasping in one hand as many stems as he is able, and with the other he pushes a pointed stick obliquely with a see-saw motion into the ground beneath the plants, by which he is able to pull up a lump of earth with the enclosed roots in an almost unbroken state. The earth is then shaken from the roots, which are subsequently carefully dried by exposure to sunshine; and then broken up into pieces a few inches in length; and finally packed in bales for exportation.

Ipecacuanha is principally imported from Brazil, but also, to some extent, from Carthagena.

General Characters. — Ipecacuanha is in more or less twisted and commonly unbranched pieces, which are usually from about two to four inches in length, and about the thickness of a small writing quill, or about one-fifth of an inch in diameter, but smaller pieces may frequently be found. It consists of a thin central whitish woody axis destitute of pith, and nearly inert; and a thick cortical or active portion, which is brownish, greyish-brown, or blackish- or reddish-brown, and marked with irregular circular fissures, which give the roots the appearance of a number of closely-arranged rings strung upon a cord, and hence the name annulated ipeca-cuanha which is applied to this root, and by which it may be more especially distinguished from other kinds of ipeca-cuanha. The bark has a resinous fracture, and the fractured surface has a semi-transparent somewhat waxy appearance. The taste is bitter and somewhat acrid; and the odour faint and peculiar, but it affects some persons to such an extent, that they cannot remain in the same house where it is being powdered. The powder has a pale brown colour.

Besides the ipecacuanha to which the above characters more particularly apply, another variety imported from Carthagena, and known as *Carthagena* or *New Granada Ipecacuanha*, is sometimes imported. It is known by its larger size, less annulated appearance, and by the radiated character of its woody axis in consequence of the large development of its medullary rays.

Adulterations.—In this country ipecacuanha is not liable to any serious adulteration, but in consequence of being badly packed, it is frequently more or less injured by sea-water, and by damp, &c. In some cases, also, a varying proportion of the more slender, nearly smooth, non-annu-lated pieces of the stems of the ipecacuanha plant are mixed with the official root; these should be rejected, as they are very inferior in activity to the true roots. Powdered ipecacuanha has also been adulterated with bitter almond meal, which is readily detected by infusion in water, from the development under such circumstances of hydrocyanic acid and volatile oil of bitter almonds.

Principal Constituents.—The essential constituent of ipecacuanha is an amorphous colourless inodorous alka-loid, with a bitter taste, termed *emetine*. It exists in the proportion of about 1 per cent., almost entirely in the cortical portion, the woody axis only exhibiting traces of its existence. It also contains a peculiar glucoside acid, named *ipecacuanhic acid, starch*, and other unimportant substances.

Medicinal Properties.—Ipecacuanha is emetic in large doses ; in small doses expectorant and diaphoretic; and in intermediate doses nauseant. Locally applied to the skin, it acts as a counter-irritant. Emetine has also been em-ployed on the Continent of Europe as a substitute for ipecacuanha, but its use requires great caution, and it has no compensating advantages.

Official Preparations.

Pilula Conii Composita.	Trochisci Morphinæ et
Pulvis Ipecacuanhæ Com-	Ipecacuanhæ.
positus.	Vinum Ipecacuanhæ.
Trochisci Ipecacuanhæ.	

Pulvis Ipecacuanhæ Compositus is also used in the preparation of Pilula Ipecacuanhæ cum Scilla.

OTHER KINDS OF IPECACUANHA.

(*Not Official.*)

Besides the true official ipecacuanha, which is distinguished as annulated ipecacuanha, other kinds of emetic roots are known in Brazil under the same common name of Poaya, and some of these, as well as other spurious ipecacuanhas from other countries, sometimes find their way to Europe. As these are all very inferior to the official ipecacuanha, the more important of them will be now briefly described, in order to prevent them, from the similarity of their names, from being confounded with it.

1. *Striated Ipecacuanha.*—There are two kinds, known as the *Large Striated* or *Black Ipecacuanha*, which is obtained from *Psychotria emetica*, Linn. (Rubiaceæ) ; and *Small Striated Ipecacuanha*, which Planchon thinks is derived from a species of *Richardsonia* (Rubiaceæ). These, as their names imply, vary in size, but both are well distinguished from annulated ipecacuanha, by having contractions at somewhat distant intervals, with irregular connecting longitudinal striations or furrows, instead of annulated markings.

2. *Undulated Ipecacuanha.* — This is derived from *Richardsonia scabra*, St. Hil. (Rubiaceæ). Externally it has a deep brownish-grey colour, and presents an irregular, sinuous, knotty, or undulating outline, without any regular annulations; and when fractured it is seen to consist of a thick white starchy cortical portion, enclosing a slender flexible woody central axis.

3. *White or Woody Ipecacuanha.*—This is obtained from *Ionidium Ipecacuanha*, Vent. (Violaceæ). It is a somewhat branched non-annulated root, of a whitish or pale-yellow colour, and marked with irregular superficial longitudinal furrows; its woody portion is porous, and of a faint yellow colour.

ORDER 3.—VALERIANACEÆ.

VALERIANA OFFICINALIS, *Linn.*
Valerian.
(Bentley and Trimen's ' Medicinal Plants,' vol. ii. plate 146.)

Habitat.—Valerian is a common plant in this country. It has a very wide distribution, from Iceland, Arctic Europe and Asia to the Mediterranean, Crimea, West Asia, and Japan.

Official Part and Name.—VALERIANÆ RHIZOMA :—the dried rhizome and rootlets.

Valerianæ Rhizoma.
Valerian Rhizome.

Synonym.—Valerianæ Radix.

Cultivation and Collection.—The official rhizome is directed to be 'collected in autumn from plants growing wild or cultivated in Britain.' It is principally obtained from the cultivated plant, although this is less active than that of the wild plant, more especially that found in dry localities, but, being finer-looking, it is commonly preferred. In this country it is cultivated to a somewhat large extent in many villages near Chesterfield in Derbyshire.

General Characters.—An erect rhizome, truncate at both ends, entire or sliced ; when entire usually from about three-quarters of an inch to one and a half inch long, and about the thickness of the little finger, and giving off some short branches or tubercules, and numerous rootlets on all sides. The rootlets are slender, brittle, three or four inches long, more or less shrivelled, and like the rhizome have a dark yellowish-brown colour externally, and are whitish within. When fresh the rhizome has a compact texture, but by keeping it becomes separated internally by transverse partitions into hollow compartments. The odour, which is developed by drying, is strong, peculiar, and disagreeable; and the taste unpleasant, camphoraceous, and somewhat bitter.

Principal Constituents.—Valerian rhizome owes its properties to a *volatile oil* and *valerianic acid,* more especially the former ; both of these are yielded by it when it is distilled with water, and hence their presence under such circumstances is given as a test of valerian rhizome in the British Pharmacopœia. The proportion of volatile oil obtainable varies much in different samples, that is, from ½ to 2 per cent.; more being obtained from plants growing in dry localities than in damp ones, or from cultivated specimens.

Medicinal Properties.—Stimulant, antispasmodic, and nervine tonic, but inferior to asafœtida as an antispasmodic. The volatile oil is also a good form of administration.

Official Preparations.

Infusum Valerianæ. | Tinctura Valerianæ.
Tinctura Valerianæ Ammoniata.

ORDER 4.—COMPOSITÆ.

1. ANACYCLUS PYRETHRUM, *DC.*

Pellitory. Pellitory of Spain.

(Bentley and Trimen's ' Medicinal Plants,' vol. iii. plate 151.)

Habitat. — Although commonly termed Pellitory of Spain, this plant does not grow wild in any part of Europe, but is confined to Northern Africa, more especially Algeria.

Official Part and Name.—PYRETHRI RADIX :—the dried root.

Pyrethri Radix.

Pellitory Root.

Commerce. — Pellitory root is chiefly obtained from Algeria by way of Oran, but some comes from Algiers.

General Characters.—Pellitory root is in sub-cylindrical or somewhat tapering pieces, which are unbranched, and usually about the length and thickness of the little finger, but vary-ing from about two to four inches in length, and from a half to three-quarters of an inch in thickness. The pieces are straight or somewhat curved, sometimes crowned at the summit by leaf-remains, and occasionally give off from their sides a few hair-like rootlets ; they are covered by a thickish brown longitudinally shrivelled bark, which is studded with dark-coloured receptacles of resin (*fig.* 32). Pellitory root has a close fracture of a yellowish-brown colour, the fractured surface presenting a radiated appearance from the large size of the medullary rays (*fig.* 32), and is marked by numerous shining dark-coloured resin receptacles ; but there is no trace of pith. It is inodorous ; but when chewed causes a burning and

Fig. 32.— Transverse section of Pellitory Root (*Pyrethri Radix*), magnified three diameters. (After Maisch.)

pricking sensation over the whole mouth and throat, followed by a copious flow of saliva.

Principal Constituents.—According to Buchheim, the active constituent is an alkaloid, *pyrethrine.* The root is also said to contain a *brown acrid resin,* an *acrid brown fixed oil,* and a *yellow acrid oil,* the former of which is more generally regarded as the chief active principle of pellitory root.

Medicinal Properties.—Pellitory root is not used inter-nally ; but it acts as a powerful local irritant and stimu-lant, causing a profuse flow of saliva when chewed, and hence termed a sialagogue.

Official Preparation.—Tinctura Pyrethri.

2. ANTHEMIS NOBILIS, *Linn.*

Chamomile. Roman Chamomile.

(Bentley and Trimen's ' Medicinal Plants,' vol. iii. plate 154.)

Habitat.—It is a rather common wild plant in the South of England, but becomes rarer in the north, and extends to Ireland, but is not thought to be a native of Scotland, unless perhaps in the Western Isles. It is also found in the west and central parts of France, and in Spain, Italy, Portugal, and Dalmatia; and is a doubtful native of Germany and Western Russia.

Official Parts or Products and Names.—1. ANTHEMIDIS FLORES :—the dried single and double flower-heads or capitula. 2. OLEUM ANTHEMIDIS :—the oil distilled in Britain from chamomile flowers.

1. Anthemidis Flores.

Chamomile Flowers.

Cultivation and Commerce.—Chamomile flowers are directed in the pharmacopœia to be obtained from cultivated plants. The most esteemed chamomile flowers are those of home cultivation, and are chiefly obtained from Mitcham. Some are also imported from abroad, especially from Germany, being largely grown in Saxony ; but likewise from France and Belgium. No commercial chamomile flowers are obtained from wild plants.

General Characters.—The single chamomile flowers of commerce are those in which the capitula have some yellow tubular central or disk florets, surrounded also by a variable number of whitish ligulate ray florets ; the double flowers are those in which all or nearly all the florets are white and ligulate. The distinction, therefore, between single and double commercial chamomile flowers is to a certain extent arbitrary. The florets are in both varieties surrounded by an involucre formed of imbricated bracts (*fig.* 33—1, A, *a* ; and 1, B, *a*), which are scarious at their margins ; the entire capitula being somewhat globular in form, and from half to three-quarters of an inch across. The receptacle in all cases is solid, conical (*fig.* 33—1, B, *b*), and densely covered with chaffy-looking scales or bractlets (*fig.* 33—1, A, *b*). Both kinds of chamomiles, but more especially the single, have a strong pleasant aromatic odour, and a very bitter taste.

Adulterations. — Chamomile flowers are not subject to any systematic adulteration in this country, but their characters ought to be carefully noted, otherwise they may be confounded with the capitula of other Compositæ; and in rare cases the flower-heads of a cultivated variety of *Chrysanthemum Parthenium*, Pers., have been detected mixed with them. These have been fully described by the author (*Pharm. Journ.* 2nd ser. vol.

Fig. 33.· 1. A. A flower-head of *Anthemis nobilis*, with the florets removed, showing the involucre, *a*, and scaly receptacle, *b*. 1. B. A flower-head of the same with florets and scales (bractlets) removed, showing involucre, *a*, and conical receptacle, *b*. 2. A. A flower-head of *Chrysanthemum Parthenium*, with the florets removed, showing the involucre, *a*, and scaly receptacle, *b*. 2. B. Flower-head of the same with florets and scales removed, showing involucre, *a*, and convex receptacle, *b*.

i. page 447, with figures) ; they are readily distinguished by their convex or nearly flat receptacle (*fig.* 33, 2, B, *b*), by having fewer scales or bractlets (*fig.* 33, 2, A, *b*), and by these being of a less membranous nature. Feverfew flowers, as they are commonly termed, have also a peculiar and disagreeable odour.

Principal Constituents.—*Volatile oil* (*see* Oleum Anthemidis), and a *bitter principle* called *anthemic acid*, which requires further investigation.

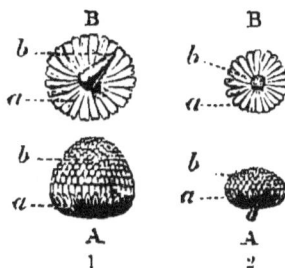

2. Oleum Anthemidis.

Oil of Chamomile.

Production.—The official oil of chamomile is directed to be distilled in Britain from chamomile flowers; but at Mitcham it is commonly obtained by distillation from the entire plant after the best flowers have been gathered.

General Characters.—When freshly distilled from the flowers it is pale blue or greenish-blue, but gradually becomes yellowish-brown; it has the peculiar aromatic odour and taste of chamomile flowers. Its specific gravity is said to be about 0·91.

Medicinal Properties of Flowers and Oil.—Chamomile flowers are stimulant, aromatic, and tonic, in moderate doses, and in large doses emetic. They are also regarded by some as antiperiodic. Flannel bags filled with hot moistened flowers are useful topical agents for the application of warmth.

Oil of chamomile possesses stimulant, stomachic, and carminative properties; and also forms a good corrective to purgative pills to prevent griping. *Dose.*—1 to 4 minims.

Official Preparations.

1. Of ANTHEMIDIS FLORES :—
    Extractum Anthemidis. | Infusum Anthemidis.
    Oleum Anthemidis.

2. Of OLEUM ANTHEMIDIS:—
    Extractum Anthemidis.

3. ARTEMISIA MARITIMA, var. STECHMANNIANA,
Besser.

Santonica.

Synonym.—ARTEMISIA PAUCIFLORA, *Weber.*

(Bentley and Trimen's ' Medicinal Plants,' vol. iii. plate 157.)

Habitat.—This plant is found in several parts of the Russian Empire, as in the lower parts of the Don and Volga, near Sarepta and Zaritzyn ; and much further to the east, in the Kirghiz desert of Russian Turkestan.

Official Parts or Products and Names.—1. SANTONICA:—the dried unexpanded flower-heads or capitula. 2. SANTO-NINUM :—a crystalline principle prepared from santonica.

1. Santonica.
Santonica.

Collection and Commerce.—Santonica is collected on the steppes or plains of the Kirghiz, in the northern parts of Russian Turkestan ; from thence it is forwarded to the fair of Nishnei-Novgorod, and reaches Europe, etc., by way of Moscow and St. Petersburg.

General Characters. — The unexpanded flower-heads forming the official santonica resemble seeds in appearance, hence they have been termed *Wormseed, Semen Santonicæ, Semen Contra, Semen Cinæ,* etc. The flower-heads are so minute that, according to ' Pharmacographia,' ' it requires about ninety to make up the weight of a grain.' In commercial specimens of santonica we frequently find the flower-heads more or less intermixed with stalks and portions of leaves. The characters are as follows :—About one-tenth of an inch in length, oblong-ovoid, obtuse, pale greenish-brown, nearly smooth ; resembling seeds in appearance, but consisting of from twelve to eighteen imbricated involucral scales with a broad thick yellowish-green

P

midrib, and enclosing three to five somewhat tubular florets. Odour, more especially when rubbed, strong, peculiar, and somewhat camphoraceous; taste bitter and camphoraceous.

Principal Constituents.—The essential constituents of santonica are from 1½ to 2 per cent. of *santonin* (*see* Santoninum), to which its anthelmintic properties would appear to be entirely due; and about 1 per cent. of a *volatile oil* which has its odour and taste.

2. Santoninum.

Santonin.

General Characters.—The characters, tests, and mode of preparation of santonin are fully given in the British Pharmacopœia, and need no further description here.

Medicinal Properties. — Both santonica and santonin, more especially the latter, possess well-marked anthelmintic properties upon the round-worm (*Ascaris lumbricoides*). Santonin has also the reputation of possessing antiperiodic properties, but on no sufficient evidence.

Official Preparations.

1. Of SANTONICA :—
Santoninum. *Dose.*—2 to 6 grains, followed by a purgative.

2. Of SANTONINUM :—
Trochisci Santonini. Each lozenge contains one grain of santonin. *Dose.*—1 to 6 lozenges.

Barbary Wormseed or Santonica.

The official kind of wormseed, which is now the only one found in commerce, has been termed *Levant* or *Alexandrian Wormseed* to distinguish it from another kind called *Barbary Wormseed*, which was formerly in use. The latter may be known by its rounded form, and by being covered with a whitish down. It is said not to yield any santonin.

4. ARNICA MONTANA, *Linn.*
Arnica. Mountain Tobacco.

(Bentley and Trimen's ' Medicinal Plants,' vol. iii. plate 158.)

Habitat.—A native of Northern and Central Europe, but not reaching the British Islands. It extends eastward through Russia to Siberia.

Official Part and Name.—ARNICÆ RHIZOMA:—the dried rhizome and rootlets.

Arnicæ Rhizoma.
Arnica Rhizome.

Synonym.—Arnicæ Radix.

Collection.—It is collected in the mountainous parts of Central and Southern Europe, generally in the spring ; and is usually imported from Germany.

General Characters.—The rhizome is sub-cylindrical in form, from one to two inches or more in length, and from about one-sixth to a quarter of an inch in diameter ; dark brown, contorted, rough from the scars of fallen leaves, some remains of which are also usually to be found at its upper end, and giving off from its under surface numerous filiform dark-brown wiry rootlets, which are frequently two or more inches in length. Odour peculiar and somewhat aromatic ; taste acrid and feebly bitter.

Adulteration.—Holmes has found as an adulteration the roots of *Geum urbanum.* These are readily known by having rootlets or their remains or scars on all sides ; and from their astringent taste.

Principal Constituents.—The principal constituents are a *volatile oil* in the proportion of about $\frac{1}{2}$ per cent. ; and a peculiar principle, supposed to be a glucoside, and on the presence of which the activity of arnica especially depends, named *arnicin.* Arnica also contains about ten per cent. of *inulin.*

P 2

Medicinal Properties.—Arnica is irritant and stimulant when given internally ; its action being more particularly directed to the spinal cord. As a local application, in the form of tincture, it is much used in bruises, sprains, etc., and is regarded as possessing a peculiar soothing and re-solvent effect, on which account it has been termed *panacea lapsorum*; but, according to Dr. Garrod, no more serviceable in such cases than spirit of the same strength.

Official Preparation.—Tinctura Arnicæ. *Dose.*—⅒ to 1 fluid drachm.

Arnica Flowers.

Arnica flowers are much used on the Continent of Europe and in the United States of America, and are probably equal, if not superior, to the rhizome in their medicinal value. They have a similar taste and odour to the official rhizome, and a yellowish colour ; and their com-position is essentially the same. For the characters of these flowers, reference must be made to botanical works.

5. TARAXACUM OFFICINALE, *Wiggers.*
Dandelion.

Synonym.—Taraxacum Dens-leonis, *Desf.*

(Bentley and Trimen's ' Medicinal Plants,' vol. iii. plate 159.)

Habitat.—A common plant throughout Great Britain, and extending over Europe, temperate Asia, Japan, and North America, and in some of its forms reaching to the Arctic regions.

Official Part and Name.—Taraxaci Radix :—the fresh and dried roots.

Taraxaci Radix.

Dandelion Root.

Collection.—It is directed in the British Pharmacopœia to be collected in the autumn from indigenous plants. But, according to the author (*Pharm. Journ.* 2nd ser. vol. i. page 402), it possesses most medicinal value at the end of February and beginning of March. It is then most bitter, and the root contains less inulin than in the autumn; hence it is inferred that as its activity depends on the presence of

FIG. 34.· *a.* Portion of the dried root of *Taraxacum officinale. b, c, d.* Transverse sections of dried Dandelion roots of different sizes.

a bitter principle, it has most medicinal activity; and moreover, as it then contains less inulin, the extract does not become opaque as in that made in the autumn, from its deposition. When dried the root is very liable to be attacked by maggots, and should not therefore be kept more than a year.

General Characters.—The root when *fresh* is frequently a foot or more in length, and half an inch or more in diameter, sub-cylindrical, but somewhat tapering below, simple or slightly branched, smooth, and yellowish-brown or brownish externally, white within. It breaks readily with a short fracture, and a milky juice exudes; the fractured surface presenting faint irregularly concentric rings, caused by the laticiferous vessels being arranged in an annular manner. It is inodorous, and has a bitter taste. In the process of drying it shrinks and loses 76 per cent. in weight. When *dried* it varies much in length and thickness, is more or less shrivelled (*fig.* 34, *a*), deeply and irregularly furrowed in a longitudinal direction, of a dark-brown or blackish colour, breaks with a short fracture, and the exposed surface shows a central porous axis of a more or less yellow colour, surrounded by a thick whitish bark, which presents a variable number, according to its size, of irregular well-marked concentric, or more or less eccentric rings (*fig.* 34, *b*, *c*, *d*), which striking character was first described by the author in a paper published in the 'Pharmaceutical Journal' in the year 1856. The root is inodorous, but has a bitter taste.

Principal Constituents.—It contains in the autumn a large quantity of *inulin*. Its active principle is generally regarded as a crystalline bitter substance, termed *taraxacin*. It also contains *resin* and another crystalline principle termed *taraxacerin*, which is insoluble in water, but soluble in alcohol. *Taraxacin* is soluble in both water and alcohol.

Medicinal Properties.—Slightly tonic, aperient, diuretic, and cholagogue.

Official Preparations.

Of the FRESH ROOT :—
　　　Extractum Taraxaci.　|　Succus Taraxaci.
Of the DRIED ROOT :—
　Decoctum Taraxaci.　|　Extractum Taraxaci Liquidum.

---

6. LACTUCA VIROSA, *Linn.*
Wild Lettuce.

(Bentley and Trimen's 'Medicinal Plants,' vol. iii. plate 160.)

Habitat.—Not uncommon in some parts of England and Scotland. It also grows throughout Western and Central Europe, and extends eastwards to Western Siberia.

Official Part and Name.—LACTUCA:—the flowering herb.

Lactuca.

Lettuce.

General Characters.—For a full description of this plant reference must be made to botanical works, but generally it may be described as having a cylindrical stem, which is prickly below, with short horizontal branches, and bearing scattered alternate horizontally arranged leaves, which are glaucous green, obovate-oblong, acute, irregularly toothed or somewhat lobed and spiny at the margins, more or less auricled and clasping at the base, glabrous generally, but the midrib whitish and spiny. The capitula are numerous, and arranged so as to form a large ovate panicled inflorescence; the corollas of the individual florets being ligulate and pale yellow. The whole herb, especially at the period of flowering, abounds in a white milky juice, which exudes wherever the plant is wounded. This juice has a bitter taste, and a strong opiate odour; by exposure to the air it hardens, and assumes a brownish colour, and then constitutes what is called *lactucarium.* This substance is, however, obtained not only from this species, but also from *Lactuca sativa*, Linn., and to some extent from *L. Scariola*, Linn., and *L. altissima*, Bieb.

Principal Constituents.—The inspissated juice termed lactucarium contains numerous substances, the more important of which are *lactucone* or *lactucerin*, *lactucin*, and *lactucic acid*, which are crystalline; and *lactucopicrin*, an amorphous substance. By distillation with water lactucarium also yields a small quantity of a *volatile oil*, to which its opiate odour is due. The active bitter principle is lactucin, but lactucopicrin is also bitter.

Medicinal Properties.—The official extract of lettuce, which takes the place of lactucarium in the British Pharmacopœia, is regarded as being somewhat narcotic, but its powers, if any, are very slight. It has also been stated to possess slightly laxative and diuretic properties.

Official Preparation.—Extractum Lactucæ. *Dose.*—5 to 15 grains.

ORDER 5.—LOBELIACEÆ.

LOBELIA INFLATA, *Linn.*

Indian Tobacco.

(Bentley and Trimen's ' Medicinal Plants,' vol. iii. plate 162.)

Habitat.—It is widely distributed throughout the eastern parts of North America, extending from Canada to the Mississippi.

Official Part and Name.—LOBELIA:—the dried flowering herb.

Lobelia.

Lobelia.

Commerce.—Lobelia is imported from the United States of America, and commonly in oblong rectangular compressed cakes or packages, weighing from half a pound to a pound each, wrapped in papers sealed at the ends, and labelled with its proper name and that of some herb-grower. It is usually grown and prepared in New Lebanon. It is also occasionally found in commerce in pieces loosely aggregated together.

General Characters.—The packages are made up of chopped pieces of the dried herb of varying lengths, which appear to have been compressed together while still moist. These pieces are yellowish-green, somewhat angular, and bear alternate, hairy, oval, irregularly toothed leaves, which are stalked below and sessile above, and from about one and a half to two inches or more long. Some flowers and fruits more or less developed are also to be found at the end of the

stalks arranged in a racemose manner.   Lobelia has a slight,
but somewhat irritating odour ; and its taste, though mild at
first, after chewing becomes burning and acrid, very similar
to that of tobacco.   Some seeds may also be frequently
seen in the fruits.   These seeds are also
found, more especially in some herb shops,
separate from the herb ; and as death has
not unfrequently occurred from their improper
use, they require a more detailed description.

Lobelia seeds (*fig.* 35) have a brown
colour; and are very minute, being on an
average only one-thirtieth of an inch long,
and one-seventy-fifth broad.   When examined by a magnifying lens they are seen to be oval or
somewhat almond-shaped, and to have their surface marked
with longitudinal and transverse ridges, with intervening
depressions, so that they have a reticulated appearance, and
resemble basket-work (*fig.* 35).   Their powder is of a brown
colour, and communicates a greasy stain to paper in which
it is wrapped.

FIG. 35.—Seed of
Lobelia inflata.

Principal Constituents.—A yellow oily volatile alkaloidal
liquid, termed *lobelina* ; traces of volatile oil, named *lobe-
lianin*; an acrid principle, called *lobelacrin*, and supposed
to be a *lobeliate of lobeline* ; and *lobelic acid.*

Medicinal Properties.—In *small doses*, expectorant and
diuretic ; in *full doses*, a nauseating emetic ; and in *excessive
doses*, a powerful acro-narcotic poison.   Its effects generally
are very similar to those of tobacco ; hence its common
name of Indian tobacco.

Official Preparations.

Tinctura Lobeliæ.   *Dose.*—10 minims to ½ fluid drachm.
Tinctura Lobeliæ Ætherea.   *Dose.*—10 minims to ½
fluid drachm.

SERIES II.—*SUPERÆ.*

ORDER I.—ERICACEÆ.

ARCTOSTAPHYLOS UVA-URSI, *Sprengel.*

Bearberry.

(Bentley and Trimen's ' Medicinal Plants,' vol. iii. plate 163.)

Habitat.—It is extensively distributed throughout the Northern Hemisphere. In the British Islands it is common in Scotland; and is also found in the North of England and Ireland.

Official Part and Name.—UVÆ URSI FOLIA :—the dried leaves.

Uvæ Ursi Folia.

Bearberry Leaves.

Collection.—Bearberry leaves should be collected for use in September or October, and in the British Pharmacopœia they are directed to be obtained from indigenous plants.

General Characters.—The dried leaves have very short stalks, are obovate or spathulate, very obtuse at the apex, coriaceous, from about half an inch to an inch in length, and from two-eighths to three-eighths of an inch in breadth. They are smooth, dark green, shining, and somewhat convex on their upper surface ; paler-coloured and minutely reticulated beneath ; and have entire and very slightly revolute margins. They have a very astringent and slightly bitter taste ; and when powdered, a feeble hay-like or somewhat tea-like odour.

Principal Constituents.—Several principles have been indicated as constituents of bearberry leaves: namely, *tannic acid, gallic acid*, a bitter crystalline neutral substance, named *arbutin*; an amorphous bitter principle, termed *ericolin*; and a colourless tasteless neutral crystalline body, named *ursone*. Their properties are more especially due to the presence of tannic and gallic acids, but also, probably, to some extent, to the bitter principles arbutin and ericolin. An infusion of bearberry leaves gives a bluish-black precipitate with perchloride of iron.

Medicinal Properties.—Evidently astringent ; and feebly diuretic. They are reputed to have a specific effect in certain diseases of the urinary organs.

Official Preparations.—Infusum Uvæ Ursi.

Substitutions and Adulterations.—Box leaves, which some-what resemble those of uva ursi, and, it is said, have been sometimes substituted for them, are at once known by their want of astringency, and by their infusion not giving a bluish-black precipitate with perchloride of iron.

A more common adulteration or substitution is that of the leaves of the *Red Whortleberry* or *Cowberry* (*Vaccinium Vitis-Idæa*, Linn.), another British plant belonging to the Ericaceæ. These much resemble them in outline and texture ; but are readily distinguished by their margins being distinctly revolute, and somewhat crenate towards the apex ; and by their dotted under surface. They have also but a feebly astringent taste ; and their infusion is coloured green by perchloride of iron instead of bluish-black.

ORDER 2.—SAPOTACEÆ.

DICHOPSIS GUTTA, *Bentl. & Trim.*

Gutta Percha.

Synonym.—ISONANDRA GUTTA, *Hook.*

(Bentley and Trimen's ' Medicinal Plants,' vol. iii. plate 167.)

Habitat.—It grows in the southern parts of the Malay Peninsula, in Borneo, Sumatra, and doubtless in other islands of the Malay Archipelago.

Official Product and Name.—GUTTA PERCHA :—the concrete juice of the above plant, and of several other trees of the natural order Sapotaceæ.

Gutta Percha.

Gutta Percha.

Preparation and Commerce.—In order to obtain this substance, it has been the custom of the Malays at Singapore and Penang to adopt the wasteful mode of first cutting down the trees ; then stripping off the bark, when the milky juice exudes, and is collected in some suitable and readily obtainable material, such as the concave stalk of a plantain leaf, the spathe of a palm, or a cocoa-nut shell. This milky juice readily coagulates on exposure to the air, and then forms what is called by the native name of *Gutta Percha.* This substance was formerly entirely derived from *Dichopsis Gutta*, but from the wasteful mode in which it has been obtained, this tree has been exterminated in some districts ; so that commercial gutta percha is now derived, as stated in the British Pharmacopœia, from it and several other Sapotaceous trees.

General Characters and Principal Constituents.—In pieces of a light-brown or chocolate colour, tough, somewhat flexible, plastic above 120° (48°·8 C.), insoluble in water, alcohol, alkaline solutions, or dilute acids ; but almost entirely soluble in chloroform, and entirely so in oil of turpentine, carbon disulphide, or benzol.

It consists of from about 72 to 82 per cent. of *pure gutta*, gutta being the Malay name for resin or gum ; a somewhat crystalline *white resin* ; and an amorphous *yellow resin.*

Uses.—Gutta Percha is applied by the surgeon to many useful purposes ; but it is in the arts where its applications and uses are most important ; these are not, however, within our province.

Official Preparation.

Liquor Gutta Percha, which solution is used in the preparation of Charta Sinapis.

ORDER 3.—STYRACEÆ.

STYRAX BENZOIN, *Dryander.*
Benzoin.

(Bentley and Trimen's 'Medicinal Plants,' vol. iii. plate 169.)

Habitat.—It is indigenous to Sumatra and Java. It is also found in Borneo and the Malay Peninsula, where it has been probably introduced.

Official Product and Name.—BENZOIN :—a balsamic resin obtained from the above plant; and probably from one or more other species of Styrax, *Linn.*

Benzoinum.

Benzoin.

Extraction, Commerce, and Kinds.—Benzoin is generally procured by making deep incisions in the bark of the trees yielding it, and allowing the liquid which exudes to concrete by exposure to the air. It is imported from both Siam and Sumatra, and in a great measure both kinds reach this country by way of Singapore and Penang. The ordinary *Sumatra Benzoin* is known to be derived from *S. Benzoin*; but the botanical source of the commercial variety from Sumatra, now distinguished under the name of *Storax-smelling Benjamin*, or *Penang Benjamin*, has not been definitely determined. The botanical source of *Siam Benzoin*, again, although commonly referred to *Styrax Benzoin*, has not been certainly ascertained.

General Characters and Kinds.—Both Siam and Sumatra benzoin, which are known in commerce as *Gum Benjamin*, have essentially the same characters, but the former is commonly of better quality, having a more agreeable odour, a finer appearance, and less mixed with extraneous substances, such as pieces of wood, bark, &c. Sumatra Benzoin has also a greyer tint than the Siam kind, and commonly fewer large

tears. Generally, the characters of benzoin are as follows :
In masses of various sizes, and either composed of loosely
agglutinated tears one or two inches long, which, when
broken, have at first an opaque milk-white appearance, but
become darker by exposure ; or, more commonly, the tears
are closely compacted together by a deep amber-brown,
reddish-brown, or greyish-brown, usually translucent substance. But in some specimens the broken tears are at first
more or less translucent, and the connecting material somewhat opaque. When the tears are thus of large size, if a
mass of benzoin is broken, it presents an almond-like appearance (*amygdaloid benzoin*) ; but in other cases the white
substance is small in quantity and much broken up, and the
connecting material abundant, so that the mass when
fractured resembles reddish-brown granite. Benzoin is very
brittle, softens readily by the warmth of the mouth like
mastich ; gives off when heated irritating fumes, which
commonly consist entirely of benzoic acid ; has very little
taste, but an agreeable vanilla-like odour, except in the case
of Penang benzoin, which resembles storax. It is soluble
in rectified spirit, and in solution of potash.

Siam benzoin was formerly imported, in some cases, in
separated tears, but this kind is now rarely or never seen in
commerce.

Principal Constituents.—The essential constituents are
amorphous resins and *benzoic acid*, and in many samples *cinnamic acid* is also found. According to some the latter acid
is only present in Siam and Penang benzoin. Benzoic acid
is official, and exists in different samples in proportions
varying from 14 to more than 20 per cent. ; it is said to be
more abundant in the dark translucent portions of benzoin
than in the white tears. The amorphous resins form the
chief portion of all kinds of benzoin ; there are at least
three varieties of these resins, but all have acid properties,
and are soluble in alcohol, although they differ in some
degree in their relations to other solvents.

Medicinal Properties.—Benzoin was formerly much used for its stimulant and expectorant properties ; it is now, however, very little employed internally. But the official compound tincture of benzoin, under the name of *Friar's balsam,* is still a favourite popular stimulant application to wounds and old ulcers. *Benzoic acid* has stimulant, expectorant, and diuretic properties ; and is also a valuable antiseptic.

Benzoin has the property of rendering fatty matters less prone to rancidity, hence it is employed as an ingredient in some official preparations for that purpose.

Official Preparations.

Acidum Benzoicum.	Tinctura Benzoini Composita.
Adeps Benzoatus.	Unguentum Cetacei.

Acidum Benzoicum is also an ingredient in Tinctura Camphoræ Composita, Tinctura Opii Ammoniata, and Trochisci Acidi Benzoici ; and is likewise used for preparing Ammonii Benzoas. Adeps Benzoatus is also employed in the preparation of numerous Unguenta.

Series III.—*DICARPIÆ OR BICARPELLATÆ.*

Order 1.—OLEACEÆ.

1. FRAXINUS ORNUS, *Linn.*

Manna Ash.

(Bentley and Trimen's ' Medicinal Plants,' vol. iii. plate 170.)

Habitat.—Native of South-Eastern Europe and of Asia Minor, and extending westward to Corsica and Eastern Spain.

Official Product and Name.—MANNA :—A concrete saccharine exudation, obtained by making transverse incisions in the stems of cultivated trees of Fraxinus Ornus, *Linn.*

Manna.

Manna.

Collection and Commerce.—Manna is at the present time collected solely in Sicily, where for that purpose the Manna Ash trees are cultivated in plantations called *frassinetti.* The manna is obtained in the months of July and August, by making daily, while dry weather lasts, a transverse incision from below upwards through the bark to the wood, when the manna exudes as a clear juice, and soon concretes on the stem, or on other substances placed for that purpose. The finest commercial manna, termed *flake manna,* is alone official ; it is obtained in the height of the season, and from the upper incisions. The smaller and inferior pieces of manna which are derived from the lower incisions, and by scraping the stems after the finer flaky pieces have been removed, are known in English commerce as *Small Manna* or *Tolfa Manna.*

General Characters.—The *official flake manna* occurs in more or less stalactitic-looking pieces of varying lengths and thickness. These pieces are flattish or somewhat grooved or concave on their inner surface, of a yellowish-brown colour, and frequently marked with adhering impurities; externally they are nearly white when fresh, and irregularly convex or stalactitic. Manna is crisp, brittle, porous, crystalline in structure ; has a faint honey-like odour ; and a sweet taste like honey, combined with a slight bitterness and acridity. It is readily soluble at ordinary temperatures in about six parts of water. It contains about 10 per cent. of moisture.

Principal Constituents.—The principal constituent of manna is *mannite* or *manna-sugar,* in the proportion, in the best kinds, of from about 60 to more than 80 per cent. Mannite occurs in colourless shining prismatic crystals, and is soluble in about six parts of cold water, and this solution does not undergo vinous fermentation in contact with yeast.

Manna also contains a varying proportion of sugar—*dextro-glucose* probably—the inferior mannas having the most; a very small quantity of a *red-brown resin*; and *fraxin*, to which the fluorescence of its aqueous or alcoholic solution is due. Mannite appears to be the essential laxative constituent of manna.

Medicinal Properties.—Manna is a mild laxative. Mannite has similar properties.

2. OLEA EUROPÆA, *Linn.*

Olive.

(Bentley and Trimen's ' Medicinal Plants,' vol. iii. plate 172.)

Habitat.—Its native country appears to be Asia Minor and Syria, but it has been cultivated from very early times in parts of the Mediterranean region.

Official Product and Name.—OLEUM OLIVÆ :—the oil expressed from the ripe fruit.

Oleum Olivæ.

Olive Oil.

Production and Kinds.—Olive oil is obtained in the South of France, Spain, and Italy, by submitting the ripe fruits or drupes, which contain about 70 per cent. of oil in their pulpy portion or sarcocarp, to moderate pressure. The fruits are collected from cultivated plants. The mode of obtaining the oil varies in some particulars in different countries and districts, which is one cause of the different qualities obtained. The finest oil, known as *virgin oil*, is that which is derived by submitting the crushed fresh fruit to moderate pressure without heat ; this oil has a greenish tint. An inferior quality of oil is obtained by moistening the marc or crushed cake left after the finest oil has been extracted, with hot water, and again submitting it to increased pressure ; and still more inferior oils are subsequently derived by pressing the residues.   Some oil, again, which is termed by the French *Huile fermentée*, is obtained by pressing the fruits after they have been gathered some time, and placed in heaps, and undergone a kind of fermentation.   In this way the amount of oil is increased in quantity, but much deteriorated in quality, speedily becoming rancid.

Various kinds of oil are known in commerce, as Provence oil, Florence, Gallipoli, Sicily, &c.   Provence oil is commonly regarded as the best.

General Characters.—Olive oil, known also as *salad oil* or *sweet oil*, is an unctuous liquid of a pale yellow or greenish-yellow colour, with a very faint agreeable odour, and a bland oleaginous taste, ultimately leaving a very slight acridity.   Its specific gravity is on an average, at 63° Fahr., about o·916° ; it is but slightly soluble in alcohol, but soluble in about twice its volume of ether.   At about 36° Fahr. (2·2° C.) it begins to congeal, and at about 21° it separates entirely into two portions: one fluid, termed *olein*, constituting about 72 per cent. of the whole ; and the other solid and crystalline, commonly called *margarin*, but in reality consisting of several solid fatty substances, in the proportion of about 28 per cent.   The above characters only apply to the finer oils.

Adulterations.—Various less expensive oils have been and are mixed with the finer kinds of olive oil, and their detection, except when in large proportion, is attended with great difficulty.   For details of these methods of ascertaining the purity of olive oil, we must refer more particularly to special treatises on chemistry.

Principal Constituents.—As mentioned above, olive oil is composed essentially of *olein* or *triolein*, and solid fatty bodies commonly known as *margarin*.   The former, so far as is known, is identical with the liquid portion of all non-drying oils ; and the latter has now been ascertained to be a mixture of *palmitin* with other substances which are compounds of fatty acids and the official glycerine. *Choles-terin* is also present to a slight extent in olive oil.

Medicinal Properties.—Demulcent and mildly laxative ; also as an antidote in cases of poisoning, where it acts mechanically by preventing absorption. Its chief use is, however, externally in skin diseases ; in burns and scalds to protect the surface from the action of the air ; and as an emollient vehicle for liniments and other external applications.

Official Preparations.

Charta Epispastica.	Linimentum Calcis.
Emplastrum Ammoniaci cum Hydrargyro.	Linimentum Camphoræ.
	Sapo Durus.
Emplastrum Hydrargyri.	Sapo Mollis.
Emplastrum Picis.	Unguentum Cantharidis.
Emplastrum Plumbi.	Unguentum Hydrargyri Compositum.
Emplastrum Saponis Fuscum.	
Enema Magnesii Sulphatis.	Unguentum Hydrargyri Nitratis.
Linimentum Ammoniæ.	Unguentum Veratrinæ.

Emplastrum Plumbi is also used in the preparation of six other Emplastra ; Linimentum Camphoræ in three other Linimenta ; and Unguentum Hydrargyri Nitratis in the preparation of Unguentum Hydrargyri Nitratis Dilutum. Sapo Mollis is also employed in the preparation of Linimentum Terebinthinæ ; and Sapo Durus is used in several preparations.

Order 2.—ASCLEPIADACEÆ.

HEMIDESMUS INDICUS, *R. Brown.*

Indian Sarsaparilla.

(Bentley and Trimen's ' Medicinal Plants,' vol. iii. plate 174.)

Habitat.—It is a common plant throughout the Indian Peninsula and in Ceylon.

Official Part and Name.—Hemidesmi Radix :—the dried root.

Hemidesmi Radix.

Hemidesmus Root.

Commerce.—It is imported from India, where it is known as *Nunnari root.* It is also commonly termed *Indian Sarsaparilla.*

General Characters.—In somewhat cylindrical more or less twisted pieces, which are usually six inches or more in length, and from one-fifth to one-half of an inch in thickness. The roots are unbranched or provided with but few rootlets, furrowed longitudinally, and frequently deeply cracked in an annular manner. They consist of a thin cortical portion, which is covered externally by a dark-brown or yellowish-brown easily separable corky layer ; and of a yellowish central woody axis, which is separated from the cortical portion by the cambium layer in the form of a dark wavy ring. They have a fragrant odour, suggestive of that of melilot or tonquin bean ; and a pleasant sweetish very slightly acrid taste.

In commercial specimens portions of the aerial stems are frequently attached to the roots ; these should be carefully separated, as they are very inferior in their properties.

Principal Constituents.—The properties are stated to be due to a *stearoptene*, which is probably the same principle as was formerly termed *similaspcric* or *hemidesmic acid.*

Medicinal Properties.—Similar to sarsaparilla, being regarded as alterative, tonic, diuretic, and diaphoretic.

Official Preparation.—Syrupus Hemidesmi.

ORDER 3.—LOGANIACEÆ.

1. STRYCHNOS NUX-VOMICA, *Linn.*

Nux Vomica.

(Bentley and Trimen's 'Medicinal Plants,' vol. iii. plate 178.)

Habitat.—It is a common plant in many parts of Southern India ; it also grows in Ceylon, Burmah, Cochin China, Java, and Northern Australia.

Official Parts or Products and Names. — 1. NUX VOMICA :—the seeds. 2. STRYCHNINA:—an alkaloid prepared from Nux Vomica.

1. Nux Vomica.

Nux Vomica.

Commerce and Varieties.—Nux vomica is imported from the East Indies, and several kinds are distinguished in the London markets under the names of Bombay, Cochin, Madras, Ceylon, &c. The Bombay is the finest kind, and has been shown by Dunstan and Short to contain a larger percentage of alkaloids than the other commercial varieties.

General Characters.—Nux vomica seeds (*fig.* 36, *a, b*) are rounded in outline, from about seven-eighths of an inch to more than an inch in diameter, and on an average nearly a quarter of an inch thick. They are nearly flat on both surfaces, or more commonly concavo-convex, or in some cases more or less bent or irregular in form; marked on one surface (the concave, when the seeds are not curved or bent) by a roundish central scar or hilum (*fig.* 36, *d*), from which a more or less projecting line (*raphe*), *e*, extends to the margin, where it ends in a slight prominence, *c*, the margin generally being rounded or somewhat keeled. The colour of the seed-coat or testa is ash-grey or yellowish-grey-green, and of a glistening appearance from being thickly covered with very short satiny adpressed silky hairs. The seeds are very hard and compact, and powdered with difficulty, the powder being of a yellowish-grey colour. Internally the seeds present a mass of horny translucent albumen, which readily splits into two

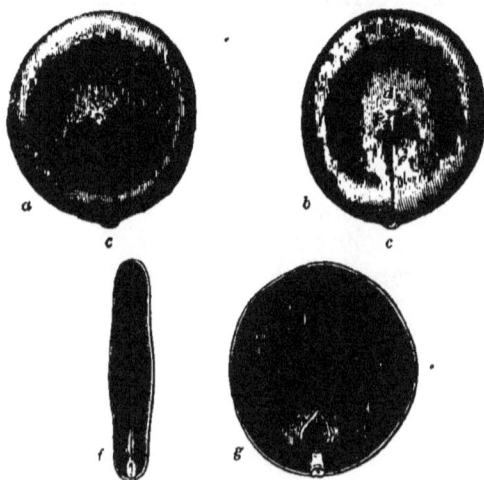

FIG. 36.—*a, b.* Seeds of *Strychnos Nux-vomica.*—*a.* Convex or dorsal surface of a seed. *b.* Concave or ventral surface. *c, c.* Prominence where the raphe ends, and marked internally by the chalaza. *d.* Hilum. *e.* Raphe. *f.* Transverse section of the seed, showing the bipartite albumen and the embryo. *g.* Vertical section, showing one-half of the albumen and the embryo.

halves, and between which, at the edge of the seed, is the embryo, with its thick club-shaped radicle, and two thin leaf-like, five- to seven-veined, somewhat heart-shaped, acute cotyledons (*fig.* 36, *f, g*). Nux vomica has no odour, but an extremely bitter taste.

Principal Constituents.—The essential constituents are two crystalline alkaloids, *strychnine* and *brucine*, combined it is said, with a peculiar acid, called *strychnic* or *igasuric acid.* A third crystalline alkaloid, named *igasurine*, has also

been indicated, but its presence has not been demonstrated
with certainty.   The total amount of strychnine and brucine,
as found by Dunstan and Short, varies in the different kinds
of nux vomica as follows :—Bombay, 3·90 per cent. ;
Cochin, 3·60 per cent. ; Madras, 3·15 per cent.   Both
strychnine and brucine are virulent poisons ; but, so far as is
known, the former is far the most important in a medicinal
point of view, and is alone official.

2. Strychnina.

Strychnine.

Synonym.—Strychnia.

General Characters and Tests.—These, as given in the
British Pharmacopœia, are as follows :—In right square octa-
hedrons or prisms colourless and inodorous; sparingly soluble
in water, but communicating to it an intensely bitter taste ;
soluble in boiling rectified spirit, and in chloroform, but not
in absolute alcohol or in ether.   Pure sulphuric acid forms
with it a colourless solution, which on the addition of bichromate of potassium acquires an intensely violet hue, speedily
passing through red to yellow.   Not coloured by nitric
acid ; leaves no ash when burned with free access of air.

Medicinal Properties.—Both strychnine and nux vomica
have powerful nervine-tonic and stimulant properties, and
are valuable remedies ; but as in improper doses they are
very poisonous, they and all their preparations should be
used cautiously, and their effects carefully watched.

Official Preparations.

1. Of Nux Vomica : — Extractum Nucis Vomicæ.
Dose.—¼ to 1 grain.   This is employed in the preparation
of Tinctura Nucis Vomicæ.   *Dose.*—10 to 20 minims.
Strychnina.   *Dose.*—$\frac{1}{30}$ to $\frac{1}{12}$ grain.

2. Of Strychnina:—Liquor Strychninæ Hydrochloratis.
Dose.—5 to 10 minims.

2. STRYCHNOS IGNATII, *Bergius.*

St. Ignatius' Bean.

Synonym.—IGNATIANA PHILIPINNICA, *Lour.*

(Bentley and Trimen's ' Medicinal Plants,' vol. iii. plate 179.)

Habitat.—Native of the Philippine Islands ; it has been introduced into Cochin China.

Part Used and Name.—IGNATII SEMINA :—the seeds. In the United States Pharmacopœia they are official under the name of Ignatia.

(*Not Official.*)

Ignatii Semina.

St. Ignatius' Beans.

Synonym.—Ignatia.

Commerce.—The supply of these seeds is very irregular. They are derived from the Philippine Islands.

General Characters.—They are about an inch or some what more long, and from a half to three-quarters of an inch broad ; oblong-ovoid in form, but presenting a very irregular surface, being rounded on one side, and presenting three or more faces with blunt angles on the other, and marked at one end by a small hilum. When fresh they are covered with short adpressed hairs of a yellowish colour; but as seen in commerce these hairs are more or less detached, and the surface is dull-grey, brownish, or blackish. They are very hard and compact, and broken with difficulty, and are then seen to consist essentially of a horny brownish somewhat translucent albumen, which, like that of nux vomica, is divided into two halves, between which at one end the embryo is placed. They are inodorous, but have a very bitter taste.

Principal Constituents.—Essentially the same as nux vomica *strychnine* and *brucine* ; the proportion of the former is stated to be 1·43 per cent., and that of the latter 0·5 per cent., but they require further examination.

Medicinal Properties.—The same as nux vomica, and are applicable therefore to the same purposes medicinally.

Preparations.

In the United States Pharmacopœia there are two preparations, namely, Abstractum Ignatiæ, and Tinctura Ignatiæ, which, like those of nux vomica, should be used with much caution, and their effects carefully watched.

---

3. SPIGELIA MARILANDICA, *Linn*
Indian Pink Root.

(Bentley and Trimen's 'Medicinal Plants,' vol. iii. plate 180.)

Habitat.—Native of the Southern United States, as far north as New Jersey and Wisconsin.

Part Used and Name.—SPIGELIA :—the rhizome and rootlets. This is the name by which it is official in the United States Pharmacopœia.

(*Not Official.*)

Spigelia.

Indian Pink Root. Carolina Pink Root.

Synonym.—Spigeliæ Radix.

Commerce.—It is principally collected in the Western and South-Western States of America. It is largely used in the United States, but only to a very small extent in this country.

General Characters.—The rhizome and rootlets are the most active portions of the plant, and are alone official in the United States, but sometimes the whole plant, with its quadrangular stems, a foot or more in length, with their attached leaves, are imported.

The rhizome is usually two or more inches long, about one-eighth of an inch thick, dark-brown, bent, somewhat branched, and marked above with the cup-shaped scars of former aerial stems. From the lower surface numerous slender, brown, branched, wrinkled and furrowed, brittle rootlets arise. It has no marked odour, but a somewhat aromatic sweetish and bitter taste.

Spigelia has some resemblance to serpentary, but is readily distinguished by its almost entire want of odour.

Principal Constituents.—Its principal constituent is a *bitter acrid principle* ; but a little *volatile oil, tannic acid,* and other unimportant principles are also present.

Medicinal Properties.—Anthelmintic, and in overdoses an acro-narcotic poison.

Preparation.

The official preparation of the United States Pharmacopœia is Extractum Spigeliæ Fluidum.

4. GELSEMIUM NITIDUM, *Michaux.*

Yellow Jasmine.

Synonym.—GELSEMIUM SEMPERVIRENS, *Aiton.*

(Bentley and Trimen's ' Medicinal Plants,' vol. iii. plate 181.)

Habitat.—Native of the Southern United States of America, from Virginia southwards ; and also extending into Mexico.

Official Part and Name.— GELSEMIUM :—the dried rhizome and rootlets.

Gelsemium.

Yellow Jasmine.

Commerce.—It is collected in the Southern United States of America. It is imported in separate pieces or in more or less compressed packets.

General Characters.—In nearly cylindrical pieces from half an inch to six inches or more in length, and commonly from a quarter to three-quarters of an inch in diameter, although sometimes more than an inch, with small roots attached to, or intermixed with, the larger pieces. Externally the pieces are light yellowish-brown, somewhat rough, and marked in a longitudinal direction by dark purplish bands or lines ; they are tough, and have a splintery fracture. They consist of a thin bark, which presents silky liber-fibres, and is closely attached to a pale yellow porous woody axis (which in the case of the rhizome has a small pith), and evident medullary rays ; the external surface of the woody axis is also marked with longitudinal elevations and corresponding depressions. The odour is somewhat aromatic and narcotic ; and the taste bitter, especially the bark.

Small pieces of the aerial stems are sometimes found mixed with the rhizome and rootlets ; they are known by their smoothness, purplish colour, and more especially by being hollow.

Principal Constituents.—A little *volatile oil, gelsemine, gelsemic acid,* and *resin.* The essential active principle is *gelsemine,* which is an amorphous alkaloid, without odour, but with a very bitter taste. It is but little soluble in water, but soluble in ether, alcohol, and chloroform.

Medicinal Properties.—It appears to resemble in some degree that of hemlock, its action being essentially on the nervous system, on which it produces a sedative effect. In overdoses it is a powerful poison. It is also said to possess febrifugal and antispasmodic properties. *Dose.*—5 to 30 grains.

Official Preparations.

Extractum Gelsemii Alcoholicum. *Dose.*—½ to 2 grains.
Tinctura Gelsemii. *Dose.*—5 to 20 minims.

ORDER 4.—GENTIANACEÆ.

1. GENTIANA LUTEA, *Linn.*
Yellow Gentian.

(Bentley and Trimen's ' Medicinal Plants,' vol. iii. plate 182.)

Habitat.—Native of the alpine and sub-alpine pastures of Central and Southern Europe, but not reaching the northern countries of the Continent nor the British Islands.

Official Part and Name.—GENTIANÆ RADIX :—the dried root.

Gentianæ Radix.
Gentian Root.

Collection and Commerce.—It is collected and dried in the mountainous districts of Central and Southern Europe. It is chiefly imported from Germany, but some is also obtained from Marseilles.

General Characters.—In more or less cylindrical pieces or longitudinal slices, or when the crown or base of the root is present, then enlarged and conical above, and covered with the scaly remains of the leaves ; from a few inches to a foot or more in length, and from about half an inch to an inch thick. Externally the pieces have a deep yellowish-brown colour, and when obtained from the upper part of the root, they are closely wrinkled in an annular manner, and all are marked with irregular longitudinal furrows ; they are yellowish or reddish-yellow within. They are tough and somewhat flexible when moist, and brittle when dry. When cut transversely, they are seen to consist of a thick reddish bark, separated by a dark-coloured cambium-zone from the central woody portion, which is somewhat spongy, and has no pith. The odour is peculiar, heavy, and unpleasant; and the taste at first sweetish, but ultimately very bitter.

Adulterations and Substitutions.—From carelessness in collection, gentian root is said to be frequently adulterated on the Continent with other roots and rhizomes ; but in this country such adulterations do not occur to any extent. The roots of other species of Gentian may, however, be sometimes found mixed with those of the official plant, but as these roots possess similar properties, such admixture is of no great importance. The roots thus found are derived from *Gentiana purpurea*, *G. pannonica*, and *G. punctata* ; those of *G. pannonica* being official in the Austrian Pharmacopœia. Such roots are not in all cases readily distinguishable ; but those of *G. purpurea* are darker-coloured internally, have no annular wrinkles, and the top of the root has a branched appearance. The roots of *G. pannonica* agree, generally, with those of *G. purpurea* ; while those of *G. punctata*, although marked with annular wrinkles, are paler-coloured externally.

Principal Constituents.—The essential bitter principle is a glucoside, which is soluble in water and in rectified spirit ; it is termed *gentiopicrin* or *gentian-bitter.* Other constituents are *gentianin* or *gentisin*, *pectin*, and *sugar.* It contains no starch, hence the official test, 'an infusion when cool is not coloured blue by tincture of iodine.'

Medicinal Properties.—Gentian is a pure bitter, and is regarded as a valuable stomachic tonic.

Official Preparations.

Extractum Gentianæ.	Tinctura Gentianæ Composita.
Infusum Gentianæ Compositum.	

2. OPHELIA CHIRATA, *Grisebach.*
Chiretta.

Synonym.—SWERTIA CHIRATA, *B. & H.*

(Bentley and Trimen's ' Medicinal Plants,' vol. iii. plate 183.)

Habitat.—Native of the mountainous regions of Northern India, Sikkim, Kumaon, Kasia, and especially Nepal.

Official Part and Name.—CHIRATA :—the dried plant.

Chirata.
Chiretta.

Collection.—The entire plant is collected in the mountainous parts of Northern India when the fruit begins to form ; it is then dried, and subsequently tied up into bundles with a slip of bamboo. These bundles are commonly flattish, about three feet long, and each from one and a half to two pounds in weight.

General Characters.—Root tapering, usually unbranched, but sometimes giving off a few rootlets, from two to four inches long, and sometimes half an inch thick in its broadest part. Stem usually solitary, two feet or more long, rounded below and in its middle part, but slightly quadrangular above, and repeatedly branched in a dichotomous manner ; it is from about two-tenths to three-tenths of an inch thick, smooth, orange-brown or purplish, and when cut across exhibiting a thin woody ring, enclosing a large continuous easily separable pith of a yellowish colour. Leaves opposite, sessile, and somewhat amplexicaul, arising from slightly thickened nodes, ovate, acuminate, five- to seven-ribbed. Flowers small, stalked, numerous, panicled, yellow. No odour ; but with a very bitter taste.

Substitutions.—In India, several other Gentianaceous plants belonging to the genera *Ophelia*, *Exacum*, and *Slevogtia*; and that of *Andrographis* (*Justicia*) *paniculata*, of the order Acanthaceæ ; are also known by the name of

Chiretta or *Chirayta*, and possess more or less the medicinal properties of the true drug.   Until the year 1874 none of these spurious drugs had been noticed in this country or elsewhere out of India, but at that period the author detected as a substitution of the true Chiretta, the dried plant of *Ophelia angustifolia*, Don, which is known in India as *Puharee* (hill) Chiretta (*Pharm. Journ.* 3rd ser. vol. v. p. 481). Since that time this substitution, as well as others, have not unfrequently occurred, but they may be readily detected by the characters of the official drug given above.   The more important of these substitutions, that of the spurious Chiretta just noticed, as being derived from *O. angustifolia*, is distinguished by the far less bitter taste of its infusion ; by the stem being evidently quadrangular throughout its whole length ; and by a transverse section showing a thick woody ring with a hollow centre, or presenting but traces of pith.

Principal Constituents. — Chiretta contains two bitter principles, namely, *ophelic acid*, and a neutral substance, termed *chiratin*, on the combined action of which the medicinal properties of the drug depend.

Medicinal Properties.—Chiretta is a pure and powerful bitter tonic, and may be employed in all cases where the use of gentian root is indicated.

Official Preparations.

Infusum Chiratæ. | Tinctura Chiratæ.

ORDER 5.—CONVOLVULACEÆ.

1. IPOMŒA PURGA, *Hayne.*

Synonym.—EXOGONIUM PURGA, *Benth.*

(Bentley and Trimen's ' Medicinal Plants,' vol. iii. plate 186.)

Habitat.—This plant is a native of the eastern slopes of the Mexican Andes, especially near Chiconquiaco.

Official Parts or Products and Names.—1. JALAPA :—the dried tubercules.   2. JALAPÆ RESINA :—the resin obtained by rectified spirit and distilled water from jalap.

1. Jalapa.

Jalap.

Collection and Commerce.—Jalap is collected in Mexico as follows :—the tubercules are dug up more or less all the year round, and hence one great cause of their variations in appearance and medicinal properties ; the best time for collection would be evidently in the autumn when the aerial stems have decayed.   After being dug up and washed, the tubercules are dried by the aid of fire heat, either entire when of small size, or the larger ones are more or less incised to facilitate their drying, or cut into halves or quarters. Jalap is imported from Vera Cruz.

General Characters.—When entire, the tubercules are irregularly oblong, somewhat ovoid, napiform, or rarely fusi-form, or even nearly cylindrical, varying much in size, usually from that of a hazel nut to a hen's egg, but often larger, sometimes even as large as a man's fist or more ; the larger ones are frequently incised as mentioned above, or cut into halves or quarters.   Externally they have a dark-brown colour, are more or less irregularly furrowed and wrinkled, and marked with paler coloured transverse lines or scars ; or very rarely they are nearly smooth.   Internally they are hard and compact, of a dirty yellow or brown colour of various shades, and frequently marked with darker-coloured irregular concentric or eccentric circles.   The powder is of a yellowish-grey or brown colour.   They have a faint, peculiar, and somewhat smoky odour, which is more per-ceptible when they are rubbed or powdered ; and their taste is sweet, acrid, and nauseous.

Principal Constituents.—*Starch, sugar*, and other ingre-dients ; but its medicinal activity is entirely due to a *resin*, which being official will be now described.

2. Jalapæ Resina.

Resin of Jalap.

Preparation, General Characters, and Tests.—This resin is directed to be obtained by a process given in the pharmacopœia by the aid of rectified spirit and distilled water, and its characters and tests are described as follows :—In dark brown opaque fragments, translucent at the edges, brittle, breaking with a resinous fracture, readily reduced to a pale-brown powder, sweetish in odour, acrid in the throat, easily soluble in rectified spirit, insoluble in oil of turpentine. The powder yields little or nothing to warm water, and not more than 10 per cent. to ether.

The amount of resin varies in different samples from about 12 to 21 per cent., and as indicated above by the ether test, this crude resin is in reality composed of two resinous substances : one which is *soluble in ether*, as well as in rectified spirit, which should not constitute, in any case, more than 10 per cent. of the whole, and commonly less ; and another forming by far the larger proportion which is *insoluble in ether*, but soluble in rectified spirit. This latter resin was formerly termed by Pereira and others *jalapin*, which name it should have retained ; but this latter name, as will be presently mentioned under the head of 'Woody Jalap,' is now more commonly applied to a resin which is *soluble in ether*; hence the official resin is more usually termed *convolvulin.*

Medicinal Properties.—Jalap and resin of jalap are powerful speedy and reliable drastic hydragogue purgatives, producing copious watery stools.

Official Preparations.

1. Of Jalapa :—

Extractum Jalapæ.	Resina Jalapæ. *Dose.*—2
Pulvis Jalapæ Compositus.	to 5 grains.
Pulvis Scammonii Compositus.	Tinctura Jalapæ.

2. Of Jalapæ Resina :—Pilula Scammonii Composita.

R

OTHER KINDS OF JALAP.

If adulterated, as is not unfrequent at the present day, the characters given above will readily serve to distinguish the true drug from the false. The roots of two false jalaps require, however, some notice from us ; these are *Tampico Jalap* and *Woody Jalap*.

1. *Tampico Jalap.*—This, as its name implies, is imported from Tampico ; its botanical source is *Ipomœa simulans*, Hanb. It resembles the official drug both in taste and smell ; but it may usually be known by its more shrivelled appearance, convoluted surface, smaller size, lighter weight, and more tapering form. But more especially by its resin, which is entirely or almost entirely *soluble in ether*, and therefore apparently analogous to jalapin. It has some resemblance to the root of *Aconitum ferox*, but this differs from Tampico jalap in its conical form, and by being marked here and there with the scars of broken-off rootlets (*see* page 18).

2. *Woody Jalap.*—This is also known as *Orizaba Root* ; *Light*, *Fusiform*, or *Male Jalap* ; or *Jalap Tops* or *Stalks*. It is obtained, as its name implies, from the neighbourhood of Orizaba, in Mexico ; and its botanical source is *Ipomœa orizabensis*, Ledanois. It occurs either in fusiform pieces, or in irregular angular or circular portions, which are sections of a larger root. It is readily distinguished from official jalap by its lighter colour and weight, and more woody character ; by the radiated appearance of its transverse section ; and also by its resin, which is *entirely soluble in ether*. It is this resin which is now generally known as *jalapin*, and is regarded as identical with scammony resin.

———

2. CONVOLVULUS SCAMMONIA, *Linn.*

Scammony.

(Bentley and Trimen's 'Medicinal Plants,' vol. iii. plate 187.)

Habitat.—A native of the East, and common in most parts of Asia Minor, Greece, the Crimea, and Syria.

Official Parts or Products and Names.—1. SCAMMONIÆ RADIX :—the dried root. 2. SCAMMONIÆ RESINA :—the resin obtained from scammony root or scammony by rectified spirit and distilled water. 3. SCAMMONIUM :—A gum-resinous exudation obtained by incision from the living root ; hardened in the air.

1. Scammoniæ Radix.

Scammony Root.

Collection.—The root is collected in Asia Minor and Syria.

General Characters.—Usually somewhat twisted, from one to two feet or more in length, and from one to two or three inches in diameter. It is commonly unbranched, nearly cylindrical except towards its upper end, where it is enlarged, and usually presents some remains of the slender aerial stems which it formerly bore ; more or less shrivelled, irregularly furrowed in a longitudinal direction, greyish-brown or yellowish externally, pale brown or whitish within, and when fractured small fragments of pale yellowish-brown resin may be often seen on the surface of the fracture. Odour and taste faint, somewhat resembling jalap.

Test.—Rectified spirit agitated with the powder and evaporated leaves a residue having the properties of scammony resin.

Principal Constituents.—The dried root contains from 3·5 to 6·5 per cent. of the official *resin*, which is its active constituent, and will now be described.

R 2

2. Scammoniæ Resina.

Resin of Scammony.

Preparation, General Characters, and Tests.—A process for its preparation is given in the pharmacopœia by means of rectified spirit and distilled water from either scammony root or scammony, and its characters and tests are described as follows :—In brownish translucent pieces, brittle, resinous in fracture, and of a sweet fragrant odour if prepared from the root. It cannot, alone, form an emulsion with water. Its tincture does not render the fresh-cut surface or inner part of the paring of a potato blue, thus showing the absence of guaiacum resin. Ether dissolves it entirely.

The latter test at once distinguishes it from resin of jalap. It is also soluble in rectified spirit. It has been termed *scammonin*, but is now regarded as identical with *jalapin*, already described as being obtained from the root of *Ipomœa orizabensis*.

3. Scammonium.

Scammony.

Collection, Preparation, and Commerce.—After the earth has been cleared from the root, and when the plant is in flower, it is cut two or more inches below the crown in a slanting direction, a mussel-shell being then immediately stuck into the root just below the lower part of the cut portion so as to receive the milky juice which immediately flows out. The shells are usually left till the evening (each shell containing on an average about sixty grains), when they are collected and their contents, as well as the semi-hardened juice which is scraped from the surface of the incised roots, are mixed in a leather bag or copper pot, etc. and carried home. The scammony as thus obtained is a somewhat transparent gummy-looking substance of a golden yellow-colour, and if dried off immediately it retains these characters. But commonly the peasant, instead of drying off the product directly, allows his daily gatherings to accumulate until he has obtained a pound or more ; he then softens it by exposure to sunshine, after which it is kneaded into a plastic mass, and finally dried. The scammony as thus prepared is that ordinarily met with in commerce, and known as *Virgin Scammony*, and although free from all foreign impurities, is not so good as that dried off as it is collected, for by being kept in a semi-liquid state and by exposure to heat, it has undergone a kind of fermentation, become darkened in colour, acquired a decayed cheese-like odour, and a porous or bubbly structure. This pure scammony is, however, often largely adulterated by being mixed by the peasants and by the dealers while in a half-dried state with carbonate of lime, wheat-flour, wood-ashes, sand, powdered scammony root, and numerous other substances, so that commercial scammony is of varying degrees of purity.

Scammony is chiefly collected in Asia Minor and exported from Smyrna. Formerly it was also obtained in Syria and forwarded from Aleppo, but no regular supplies are now derived from the latter source.

General Characters and Tests.—Scammony as dried off directly after its collection, in the manner described above can scarcely be called a commercial article ; but the best scammony as met with ordinarily in commerce, as just noticed, and commonly called *Virgin Scammony*, presents the following characters :—In flattish cakes or pieces of irregular form and of varying sizes; ash-grey or blackish-brown externally, and sometimes sprinkled over with a greyish-white powder. It is very brittle, and when fractured the surface is resinous-looking, shining, more or less porous or bubbly, of a uniform dark greyish-black colour, and if rubbed with the moistened finger it forms a whitish emulsion, by which it is known from resin of scammony ; and also when touched with hydrochloric acid it does not effervesce, thus indicating the absence of chalk. It is easily powdered, the powder being of an ash-grey colour, and when this is triturated with water it forms a smooth emulsion. A decoction when cool is not rendered blue by solution of iodine, thus proving the absence of starch and amylaceous substances generally. Paper moistened with an ethereal tincture of scammony should undergo no change of colour when exposed to nitrous fumes, thus showing the absence of guaiacum resin. It has a peculiar odour, resembling old cheese; and when chewed it causes in a short time a slight pricking sensation in the back of the throat. Ether should remove not less than 75 per cent. of resin; and what remains is chiefly soluble gum, with a little moisture.

Medicinal Properties.—Scammony and scammony resin closely resemble jalap and jalap resin in their effects, which are those of a powerful drastic hydragogue cathartic; but on account of its irritant action it is generally given in combination with other purgatives. It is also used as a vermifuge.

Official Preparations.

1. Of Scammoniæ Radix :—Resina Scammoniæ. *Dose.* —3 to 8 grains.

2. Of Scammoniæ Resina :—

Confectio Scammonii.	Pilula Colocynthidis Composita.
Extractum Colocynthidis Compositum.	Pilula Scammonii Composita.
	Pulvis Scammonii Compositus.

Pilula Colocynthidis Composita is also used in the preparation of Pilula Colocynthidis et Hyoscyami.

3. Of Scammonium :—

Mistura Scammonii. | Resina Scammoniæ

ORDER 6.—SOLANACEÆ.

1. CAPSICUM FASTIGIATUM, *Blume*.

East Indian Capsicum.　Guinea Pepper.

(Bentley and Trimen's ' Medicinal Plants,' vol. iii. plate 188.)

Habitat.—It is found apparently wild in Southern India ; and also abundantly in Java and other parts of the Eastern Archipelago, though probably in all these parts originally introduced from America.　It is extensively cultivated in Tropical Africa and America.

Official Part and Name.—CAPSICI FRUCTUS :—the dried ripe fruit.

1. Capsici Fructus.

Capsicum Fruit.

Commerce.—In commerce the fruits of this and other species and varieties are known as Pod Pepper, and of these, two are found in British commerce, namely, the *official fruit* now under description, and generally distinguished as *Guinea Pepper* ; and that derived from *Capsicum annuum*, Linn., which is more commonly known in this country as *Capsicums* ; or sometimes, as well as the former, as *Chillies.*　This latter is *not official.*

The official fruit is principally imported from Sierra Leone, Natal, and Zanzibar; but also from Bombay and other parts of the East, as Penang and Pegu.

General Characters.—Capsicum fruits vary from about a half to three-quarters of an inch in length, from about one-fifth to a quarter of an inch in diameter, and in some cases have the remains of the calyx at their base ; they are oblong-conical or ovoid-oblong in form ; blunt-pointed ; somewhat compressed ; and composed of a more or less shrivelled smooth shining thin translucent brittle leathery pericarp, of a dull orange-red colour, enclosing about eighteen roundish or ovoid-flat seeds, which are about

one-eighth of an inch in diameter. Taste of both pericarp
and seeds intensely pungent ; odour peculiar and pungent.

The powdered fruits of *Capsicum fastigiatum* are the
principal source of *cayenne pepper*; but this is also derived,
to some extent at least, from the fruits of *C. annuum*, and
of probably other capsicum fruits.

Principal Constituents.—The active principle was formerly stated to be an acrid oleo-resinous liquid obtained by
digesting an alcoholic extract in ether, and evaporating the
ethereal solution ; it is commonly named *capsicin*. But the
investigations of Dr. Thresh have shown that this is a complex body, and that the real active principle is a crystalline
colourless substance, which he obtained in very minute pro·
portion, and has named *capsaicin*. This is a very acrid
volatile principle, which is soluble in alcohol, ether, and
the fixed oils.

Medicinal Properties.—Internally powerfully stimulant;
but it is principally used as a local stimulant in the form
of a gargle, and also as a rubefacient in the form of a liniment.

Official Preparation.—Tinctura Capsici.

2. Capsicum annuum.

(*Not Official.*)

The fruit of *Capsicum annuum* varies much in form and
size, but is always much longer and broader than the *official
kind*—that is, from two to three or more inches in length,
and from about a half to three-quarters of an inch in
diameter.   It is oblong-conical in form, tapering at the
apex, and having at its base the remains of the calyx and
pedicel; and consisting of a scarlet or deep orange-red,
smooth, somewhat flattened and shrivelled, leathery pericarp,
enclosing numerous flattened yellowish-white seeds, which are
somewhat larger than those of *Capsicum fastigiatum*. There
is no marked odour, but the taste is hot and pungent.

2. SOLANUM DULCAMARA, *Linn.*

Bitter-Sweet.   Woody Nightshade.

(Bentley and Trimen's ' Medicinal Plants,' vol. iii. plate 190.)

Habitat.—Common in England, but more scarce in Scotland. It is also found throughout Europe, except the extreme north, and through Western Asia to North-West India and China. It is also wild in Northern Africa and Asia Minor, and is naturalised in many parts of North America.

Part Used and Name.—DULCAMARA :—the dried young branches.

(*Not Official.*)

Dulcamara.

Dulcamara.

Synonym.—Stipes Dulcamaræ.

Collection.—The young branches should be collected from indigenous plants, either late in the year when they have shed their leaves, or early in the spring before the leaves have appeared ; and then cut into short lengths and dried.

General Characters.—The pieces are nearly cylindrical in form, faintly angular, light in weight, somewhat striated longitudinally, and about a quarter of an inch or less in diameter. They are covered by a thin bark, the corky layer of which is shining and of an ash-grey or pale greenish-brown colour, and marked with alternately-arranged leaf-scars ; beneath the corky layer the green layer (*phelloderm*) is very marked. The wood is radiate, with usually one or two concentric rings, and lined internally by the remains of the pith, but as this only partially fills the axis, the centre is hollow. They have but little odour ; their taste is bitter at first, but subsequently sweetish, hence the common name, *bitter-sweet*, which is given to the plant.

Principal Constituents. — The principal constituents appear to be a glucoside of a bitter-sweet taste, which is resolvable into *glucose* and *solanine,* and named *dulcamarin* or *dulcarin;* and an amorphous alkaloid, termed *dulcamarine,* which is also said to have a bitter-sweet taste, but it requires further examination.

Medicinal Properties. — It is said to be alterative, sedative, and somewhat diuretic ; but its use in this country is almost obsolete.

Preparations. — The preparation ordered in the British Pharmacopœia of 1867, when it was official, was Infusum Dulcamaræ, which was made by infusing one ounce of bruised dulcamara in ten fluid ounces of boiling distilled water for one hour, and then straining. In the present United States Pharmacopœia an Extractum Dulcamaræ Fluidum is ordered.

3. NICOTIANA TABACUM.
Tobacco.
(Bentley and Trimen's 'Medicinal Plants,' vol. iii. plate 191.)

Habitat. — Originally a native of some part of Central or South America, but the precise country cannot now be determined. It is not now known anywhere in a truly wild state.

Official Part and Name. — TABACI FOLIA :— the dried leaves.

Tabaci Folia.
Leaf Tobacco.

Preparation and Commerce. — Two states of tobacco are known in commerce — one termed *unmanufactured* or *leaf tobacco,* and the other *manufactured tobacco* ; the first only is official. It is obtained from the United States, and more especially Virginia, and hence it is frequently known as Virginian Tobacco.

General Characters and Tests. — The characters and

tests are thus given in the British Pharmacopœia :—Large, being sometimes more than twenty inches long ; ovate, ovate-lanceolate, or oval-oblong, acute, entire, brown, brittle, glandular-hairy ; having a characteristic odour and nauseous bitter acrid taste ; yielding, when distilled with solution of potash, an alkaline fluid, which has the peculiar odour of nicotina, and precipitates with perchloride of platinum and tincture of galls.

Principal Constituents.—The active principle is a peculiar colourless volatile oily alkaloid, named *nicotine*, which is very soluble in water, alcohol, and ether. Tobacco leaves also yield, when distilled with water, a colourless tasteless crystalline substance, with a tobacco-like odour, which has been termed *nicotianin* or *tobacco camphor*.

Medicinal Properties.—Tobacco possesses powerfully sedative and antispasmodic properties ; it is also laxative, emetic, and diuretic. Externally applied in the form of snuff it acts as a sternutatory, etc. At present it is but very little employed medicinally in consequence of its violent action. In overdoses it is an acro-narcotic poison ; and its alkaloid nicotine is a most virulent poison.

There are *no official preparations* in the present British Pharmacopœia ; but in the Pharmacopœia of 1867, a formula was given for Enema Tabaci.

---

4. DATURA STRAMONIUM, *Linn.*
Stramonium. Thorn Apple.

(Bentley and Trimen's ' Medicinal Plants,' vol. iii. plate 192.)

Habitat.—Supposed to have been originally a native of the countries bordering the Caspian or adjacent regions. It is now spread throughout the world, except in the colder temperate and arctic regions. It is not uncommon in the South of England, but scarcely naturalised.

Official Part and Name.—STRAMONII SEMINA :—the dried ripe seeds.

1. Stramonii Semina.

Stramonium Seeds.

General Characters and Tests.—About one-sixth of an inch long, somewhat reniform, flattened ; testa brownish-black, finely-pitted, wrinkled, and enclosing a much-curved embryo, surrounded by a whitish oily albumen. Inodorous when entire, but when bruised having a disagreeable odour ; their taste is bitterish. A tincture prepared by digesting the entire seeds in dilute alcohol presents a greenish fluorescent appearance, which becomes yellow on the addition of ammonia. The seeds are much more active than the leaves.

Principal Constituents.—The activity of both the seeds and leaves are due to the alkaloid *daturine*, which is in combination with *malic acid.* This alkaloid is contained in much larger proportion in the seeds than the leaves; the former only containing it, however, in the proportion of about $\frac{1}{10}$ per cent. It is a very energetic poison, and is frequently regarded as analogous to *atropine.* According to Ladenburg, daturine is a mixture of *atropine* and *hyoscyamine.* (*See* Atropina.)

Medicinal Properties.—Analogous to belladonna, being sedative, anodyne, and antispasmodic ; and in overdoses it is a powerful poison. Like belladonna, it dilates the pupil of the eye.

Official Preparations.

Extractum Stramonii. *Dose.*—$\frac{1}{4}$ to $\frac{1}{2}$ grain.
Tinctura Stramonii. *Dose.*—10 to 30 minims.

2. Stramonii Folia.

Stramonium Leaves.

(*Not Official.*)

In the British Pharmacopœia of 1867, the dried leaves were also official, and were directed to be collected from plants in flower, and cultivated in Britain.

General Characters.—Petiolate, varying much in size, the lowest and longest often as much as eight inches or more ; ovate, unequal at the base, acuminate, deeply indented at the margins, with large irregular pointed spreading teeth or lobes, nearly or quite smooth. They are thin, brittle, dull-green in colour, paler beneath ; almost inodorous, or faintly tea-like, and have a bitterish saline disagreeable taste.

Principal Constituents and Medicinal Properties.—(*See* Stramonii Semina.) The leaves of this species and those of Datura Tatula, *Linn.*, are principally employed in the form of cigarettes in asthma.

5. ATROPA BELLADONNA, *Linn.*
Belladonna. Deadly Nightshade.
(Bentley and Trimen's ' Medicinal Plants,' vol. iii. plate 193.)

Habitat.—Not uncommon in England, but rare in Scotland and Ireland. It is also a native generally of Central and Southern Europe; and is likewise found in Asia Minor and Algeria.

Official Parts or Products and Names.—1. BELLA-DONNÆ FOLIA :—the fresh leaves, with the branches to which they are attached ; also the leaves separated from the branches and carefully dried ; gathered, when the fruit has begun to form, from plants growing wild or cultivated in Britain. 2. BELLADONNÆ RADIX :—the root from plants growing wild or cultivated in Britain, and carefully dried ; or imported in a dried state from Germany. 3. ATROPINA :—an alkaloid obtained from belladonna.

1. Belladonnæ Folia.

Collection.—As directed, the leaves should be collected when the fruit has begun to form, as they are then in their most active state, for the reasons explained under ' Aconiti Folia.' (*See* page 13.)

General Characters.—The fresh leaves are alternately arranged below, but in pairs above of unequal size, all shortly stalked, from three to eight inches long, broadly ovate or somewhat oval, tapering into the petiole, acute, entire, dark-green, smooth, veiny. Their odour, when bruised, is somewhat disagreeable ; and their taste feeble, bitterish, and somewhat acrid.

The *dried leaves*, which are usually separate from the stalks, are brownish-green above, greyish beneath, but in other respects they resemble the fresh leaves except that they have no marked odour.

Test.—The expressed juice of the fresh leaves, or an infusion of the dried leaves, dropped into the eye, dilates the pupil.

Principal Constituents.—*See* ' Belladonnæ Radix.'

2. Belladonnæ Radix.

Belladonna Root.

Collection.—The roots should be collected for drying in the autumn or early spring, and from plants two to four years old. The British prepared roots are more to be depended upon than those obtained from Germany, and those about the size of the middle finger are commonly preferred to roots of larger size.

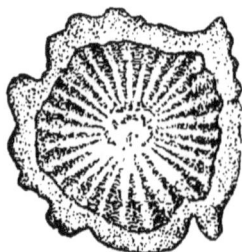

FIG. 37.—Transverse section of Belladonna Root (*Radix Belladonnæ*). (After Maisch.)

General Characters. — In rough irregular branched pieces, from one to two feet long, and from half an inch to two inches or more in diameter, and generally marked at their upper end by the hollow bases of the stems which they once bore. The roots are covered with a dirty grey or brownish integument, which is easily scraped off by the nail, when the exposed surface presents a whitish appearance. Their fracture is short, and the exposed surface is then seen to consist of a thinnish cortical portion (*fig.* 37), of a yellowish or pale-brown colour, separated by a dark line from a large central woody portion of a brownish colour, which is marked throughout by scattered darker-coloured dots, and with or without evident medullary rays—that is, the main root and large branches with, and the smaller branches without. Nearly inodorous; taste sweetish at first, but subsequently acrid and feebly bitter.

Test.—An infusion dropped into the eye dilates the pupil.

Principal Constituents.—Both leaves and root owe their activity to the alkaloid *atropine*, which, being official, is described below. (*See also* Daturine, page 252.) According to some chemists, the root also contains a second, but uncrystallisable alkaloid, termed *belladonnine*; which, as suggested by Ladenburg, is probably identical with *hyoscyamine*.

3. Atropina.

Atropine.

Synonym.—Atropia.

Preparation.—Atropine is contained in both the leaves and root, but, according to most chemists, in larger proportion in the latter, from whence it is directed to be obtained in the British Pharmacopœia by a process there given. According to Gerrard, wild belladonna contains more alkaloid than the cultivated, and the leaves more than the root. It is contained in variable proportions in the roots; thus, according to different chemists, from 0·25 to 0·6 per cent.; and Dunstan and Ransom found in the dried leaves from 0·15 to 0·22 per cent. The moderate-sized and young roots yield more than those which are old and thick.

General Characters and Tests.—As mentioned under 'Stramonii Semina,' atropine is sometimes regarded as analogous in composition, characters, and properties to daturine. Its characters and tests are given in the British Pharmacopœia as follows :—In colourless acicular crystals, sparingly soluble in water, more readily in alcohol and in ether. Its solution in water has an alkaline reaction, gives a citron-yellow precipitate with perchloride of gold, has a bitter taste, and powerfully dilates the pupil. It leaves no ash when burned with free access of air. It is an active poison.

Medicinal Properties.—Both the leaves and root are regarded as anodyne, sedative, antispasmodic, diuretic, stimulant, and lactifuge ; and when applied locally or taken internally, they dilate the pupil of the eye. In overdoses belladonna is a powerful poison. Atropine is scarcely employed internally, but much used in the form of solution of sulphate of atropine for subcutaneous injection ; and by the ophthalmic surgeon to produce dilatation of the pupil of the eye in operations, etc.

Official Preparations.

1. Of BELLADONNÆ FOLIA :—
Extractum Belladonnæ (fresh leaves and young branches). *Dose.*—¼ to 1 grain.
Succus Belladonnæ (fresh leaves and young branches). *Dose.*—5 to 15 minims.
Tinctura Belladonnæ (dried leaves in powder). *Dose.*— 5 to 20 minims.

2. Of BELLADONNÆ RADIX :—
Atropina. (*See below.*)
Linimentum Belladonnæ.
Extractum Belladonnæ Alcoholicum. *Dose.*—¹⁄₁₆ to ¼ grain. This is also employed in the preparation of Emplastrum Belladonnæ, and Unguentum Belladonnæ.

3. Of ATROPINA :—Atropinæ Sulphas, which is used in the preparation of Liquor Atropinæ Sulphatis, and Lamellæ Atropinæ. *Dose* of Liquor Atropinæ Sulphatis, 1 to 4 minims.—Each lamella contains $\frac{1}{5000}$ grain of sulphate of atropine.
Unguentum Atropinæ.

6. HYOSCYAMUS NIGER, *Linn.*
Henbane.
(Bentley and Trimen's 'Medicinal Plants,' vol. iii. plate 194.)

Habitat.—Not uncommon in England and Ireland, but rare in Scotland. It is found throughout Europe, except the extreme North, and is very common in the Mediterranean regions and Western Asia, extending to India and Siberia.

Official Part and Name.—HYOSCYAMI FOLIA :—the fresh leaves and flowers, with the branches to which they are attached ; also the leaves separated from the branches, and flowering tops, carefully dried. Collected from biennial plants, growing wild or cultivated in Britain, when about two-thirds of the flowers are expanded.

Hyoscyami Folia.
Henbane Leaves.

Collection.—The reasons why the leaves of plants should be collected at varying times after the flowering stage has advanced have already been described in treating of ' Aconiti Folia ' (*see* page 13). The young herbaceous branches are here, and in other cases, directed to be taken with the leaves, because they, like the leaves, are assimilating organs, and for other reasons which are fully explained by the author in the 'Pharmaceutical Journal,' 2nd ser. vol. iii. page 476.

General Characters.—Leaves varying much in length, sometimes as much as ten inches, with or without a stalk, alternately arranged on sub-cylindrical glandular-hairy branches, exstipulate. They are triangular-ovate or ovate-oblong, acute, undulated, irregularly toothed, sinuated, or pinnatifid, pale green, glandular-hairy, especially on their under surface, and with the midrib prominent below. Flowers with a bell-shaped calyx, and a large funnel-shaped corolla with a limb divided into five shallow rather unequal

S

lobes, which are straw-coloured, elegantly net-veined with purple, and with a purple throat. The fresh herb has a strong unpleasant narcotic odour, and a bitter slightly acrid taste.

When dried, the leaves become greyish-green, and much shrivelled except at the midrib which is very evident. The odour is also much diminished by drying, but their taste is more marked. The dried leaves, as seen in the pharmacies, are also generally much broken, more or less aggregated together, and have some flowers mixed with them.

Test.—The expressed juice of the fresh leaves, or an infusion of the dried leaves, when dropped into the eye, dilates the pupil.

Henbane Seeds.—These are *not official* in the British Pharmacopœia; but their characters, which are as follows, ought to be known. Small, being from about one-eighteenth to one-sixteenth of an inch broad, somewhat compressed, roundish in outline, light brown, finely pitted and reticulated; almost inodorous; but with an oily bitter somewhat acrid taste. Embryo much curved in oily albumen.

Principal Constituents.—The seeds, leaves, and herb generally, owe their properties essentially to the alkaloid *hyoscyamine*; the leaves containing from 0·042 to 0·224 per cent. It is said to be identical with duboisine, and isomeric with atropine. According to Ladenburg, it also contains a second, but uncrystallisable, alkaloid termed *hyoscine*. The seeds also contain about 25 per cent. of a *fixed oil*.

Medicinal Properties.—Henbane has essentially the same properties as belladonna, and like it dilates the pupil of the eye, whether given internally or locally applied.

Official Preparations.

Extractum Hyoscyami. This is also used in the preparation of Pilula Colocynthidis et Hyoscyami.

Succus Hyoscyami. *Dose.*—½ to 1 fluid drachm.

Tinctura Hyoscyami. *Dose.*—½ to 1 fluid drachm.

ORDER 7.—SCROPHULARIACEÆ.

DIGITALIS PURPUREA, *Linn.*

Foxglove.

(Bentley and Trimen's ' Medicinal Plants,' vol. iii. plate 195.)

Habitat.—It is a common plant in most parts of this country. It is also found in nearly all parts of Europe, except Greece and Turkey ; and occurs, although probably introduced, in the Azores and Madeira.

Official Part and Name.—DIGITALIS FOLIA :—the leaves. Collected from wild British plants of the second year's growth when about two-thirds of the flowers are expanded, and carefully dried.

Digitalis Folia.

Foxglove Leaves.

Collection.—The reasons why the leaves are directed to be obtained when about two-thirds of the flowers are expanded, have already been explained under ' Aconiti Folia.'

General Characters.—The leaves vary in length and breadth according to the part of the stem from whence they are derived : thus the lower ones are sometimes one foot or more in length, five or six inches in breadth, and are furnished with a long winged petiole ; those above the radical leaves become gradually shorter and narrower, and more shortly stalked, until they terminate in sessile leaf-like bracts. They are ovate or ovate-lanceolate, sub-acute, crenate or irregularly crenate-dentate, somewhat rugose, slightly hairy and dull green above, densely pubescent and paler beneath ; the veins on the under surface are evidently reticulated, the primary ones proceeding from a thick fleshy midrib at a very acute angle. When bruised, the fresh leaves have a disagreeable odour, but when dried this

S 2

becomes agreeable and tea-like ; they have a very bitter unpleasant taste.

Adulteration.—The dried leaves of other British plants are sometimes mixed with, or substituted for, those of foxglove leaves. The more common adulterations are, the leaves of *Inula Conyza*, DC., Ploughman's Spikenard; *Inula Helenium*, Linn., Elecampane; *Symphytum officinale*, Linn., Comfrey; and those of *Verbascum Thapsus*, Linn., Mullein. The characters given above of foxglove leaves ought readily to enable the true leaves to be distinguished from the spurious. But it may be further stated that the leaves of the first plant have nearly entire margins, and those of the second, although toothed, have the primary veins nearly at right angles to the midrib, while both are less reticulated on their under surface. The leaves of Comfrey have more hispid hairs, and their margins are entire ; and those of Mullein are at once known by being coated with branched stellate hairs, so that both surfaces present a woolly appearance.

Digitalis Seeds.—These seeds are *not official* ; their characters are as follows :—Minute, oblong or ovoid, somewhat angular, light brown in colour, deeply alveolate, with a straight embryo in the axis of albumen.

Principal Constituents.—The activity of both leaves and seeds has been ascribed to a principle called *digitalin* ; but this name has been applied by chemists to different substances. Thus the *digitalin of Nativelle* is a substance in needle-shaped crystals and of active medicinal properties. Digitalin is no longer official in this country, that of the British Pharmacopœia of 1867 was the *digitalin of Homolle*. It is also a very active substance, but not crystalline. Other principles obtained from digitalis are *digitoxin*, which is crystalline and intensely poisonous ; and an amorphous yellowish substance *toxiresin*, which is also a powerful poison.

Medicinal Properties.—Foxglove leaves possess great

potency, acting especially as a sedative of the heart's action ; but also as a diuretic. In improper doses foxglove is an acro-narcotic poison. The dose of the powder is from $\frac{1}{2}$ to $1\frac{1}{2}$ grain.

Official Preparations.

Infusum Digitalis. *Dose.*—2 to 4 fluid drachms.
Tinctura Digitalis. *Dose.*—10 to 30 minims.

ORDER 8.—LABIATÆ.

1. LAVANDULA VERA, *DC.*
Lavender.

(Bentley and Trimen's ' Medicinal Plants,' vol. iii. plate 199.)

Habitat.—It is a native of the South of France, Spain, Northern Italy, and other parts of the Mediterranean region, including North Africa, but does not reach Asia Minor.

Official Product and Name.—OLEUM LAVANDULÆ :—the oil distilled in Britain from the flowers.

General Characters of the Flowers.—The flowers when fresh have a violet colour, a strong fragrant odour, and a pleasant aromatic somewhat bitterish taste. When used as a perfume or for pharmaceutical purposes, the flowers are generally stripped from their stalks, and dried by a moderate heat. When thus prepared, they have a greyish-blue colour ; and their odour is retained and preserved for a long time. They are not usually kept by pharmacists.

Oleum Lavandulæ.
Oil of Lavender.

Preparation.—The oil is commonly distilled from the flower-stalks and flowers together ; and either when fresh or in a more or less dried state ; but the best oil, as directed in the pharmacopœia, is distilled from the flowers after they have been separated from their stalks.

The oil distilled in Britain is *alone official*, as it is very superior to foreign oil of lavender. The plants for this purpose are chiefly cultivated at Mitcham, Hitchin, and Market Deeping.

General Characters and Composition.—Pale yellow or nearly colourless, with the very fragrant odour of the flowers, and a hot bitter aromatic taste. It is levogyre, and has a specific gravity, according to Zeller, varying from 0·87 to 0·94 ; and is readily soluble in alcohol. It reddens litmus paper. It is said to be a mixture of oxygenated oils and stearoptene—which latter, according to Dumas, is identical with our official camphor.

Medicinal Properties.—Stimulant and carminative. *Dose.* —1 to 4 minims.

Official Preparations.

Linimentum Camphoræ Compositum.

Spiritus Lavandulæ.

Tinctura Lavandulæ Composita, which is also used in the preparation of Liquor Arsenicalis.

---

2. MENTHA VIRIDIS.

Spearmint. Common Garden Mint.

(Bentley and Trimen's ‘Medicinal Plants,’ vol iii. plate 202.)

Habitat.—This is the common Garden Mint, and is regarded by Bentham as not improbably a cultivated race of *Mentha sylvestris*, Linn. (Horse-mint). It is found wild or semi-wild in countries where it has been long cultivated, as in England.

Official Product and Name.—OLEUM MENTHÆ VIRIDIS : the oil distilled in Britain from fresh flowering spearmint.

Cultivation of Spearmint.—This is the common Mint cultivated in market gardens and kitchen gardens. It is also grown to a small extent at Mitcham and elsewhere ; it is chiefly sold in a dried state for culinary purposes. *It is only mentioned in the British Pharmacopœia as the source of the volatile oil*, in which its medicinal properties essentially reside.

General Characters of Spearmint.—The stem is quadrangular, erect, two to three feet high, green generally, but often tinged of a bright purple, and smooth or very slightly hairy. Leaves opposite, sessile or nearly so, lanceolate or oblong-lanceolate, two to four inches long, acute, serrate, dark green and smooth above, paler beneath, glandular and smooth, or only slightly hairy on the principal veins. Flowers in terminal verticillasters, the lower ones much separated ; corolla pale purple, naked within and without. It has a strong aromatic pleasant odour, and a warm slightly bitter taste in a fresh state ; and although these qualities are somewhat less marked in the dried herb, they are long retained.

Oleum Menthæ Viridis.

Oil of Spearmint.

Preparation and Commerce.—The oil is rarely distilled in Britain on account of its cost ; that generally in use is obtained from the United States of America and from Germany ; the latter oil is, however, said to be obtained from *Mentha aquatica*, Linn., var. *crispa*, Bentham.

General Characters and Composition.—Colourless, pale yellow, or greenish-yellow when recent, but becoming reddish by age, with the odour and taste of the herb. It is levogyre, has a specific gravity of 0·914, and is freely soluble in alcohol. According to Gladstone, it consists of a hydrocarbon almost identical with oil of turpentine mixed with an oxidised oil to which is due the peculiar odour of the plant.

Medicinal Properties.—The herb has properties similar to, but milder than, peppermint, being aromatic, carminative, mildly stimulant, and stomachic ; the oil is stimulant and carminative, and also useful as a flavouring agent. *Dose.*—1 to 4 minims.

Official Preparation.—Aqua Menthæ Viridis.

---

3. MENTHA PIPERITA, *Smith.*
Peppermint.

(Bentley and Trimen's ' Medicinal Plants,' vol. iii. plate 203.)

Official Products and Names. — 1. OLEUM MENTHÆ PIPERITÆ :—the oil distilled in Britain from fresh flowering peppermint.   2. MENTHOL :—a stearoptene obtained by cooling the oil distilled from the fresh herb of Mentha arvensis, *DC.*, vars. piperascens et glabrata ; and of Mentha piperita, *Sm.*

Cultivation of Peppermint.—It is extensively cultivated at Mitcham, Market Deeping, Hitchin, and Wisbeach, in England ; but is *only official as the source of the volatile oil*, in which its properties reside ; and as one source of menthol.

General Characters of Peppermint.—The herb has much resemblance to that of spearmint, but is readily distinguished by its leaves being all evidently stalked, by the more crowded arrangement of its verticillasters, its larger flowers, and by its more powerfully pungent taste and peculiar odour.   The dried herb has a similar taste and odour, which persists for a long time.

1. Oleum Menthæ Piperitæ.
Oil of Peppermint.

Preparation.—The oil is chiefly distilled in Britain at Mitcham and Market Deeping, and although the oils thus

derived are of varying qualities, they are regarded as very
superior to foreign oils.

General Characters and Composition.—Colourless, pale
yellow, or greenish yellow when recent, but becoming
gradually thicker and reddish by age, with the odour of
peppermint, and a strong penetrating aromatic taste, followed by a sensation of coldness in the mouth. It is levogyre, has a specific gravity varying from o·84 to o·92, and is
freely soluble in alcohol. It consists of two parts, one fluid,
the other solid; the former has not yet been fully examined;
but the latter, which is deposited when oil of peppermint
is cooled to about eight degrees below the zero of Fahrenheit, is the official menthol described below, and to it the
odour and taste of peppermint are due.

Medicinal Properties.—Peppermint possesses aromatic,
carminative, antispasmodic, and stimulant properties ; the
oil has similar qualities, and is largely used as an adjunct to
other medicines, and as a flavouring agent.

Official Preparations.

Aqua Menthæ Piperitæ.   Spiritus Menthæ Piperitæ.
Essentia Menthæ Piperitæ.   Tinctura   Chloroformi   et
Pilula Rhei Composita.   Morphinæ.

Aqua Menthæ Piperitæ is also used in the preparation of
Mistura Ferri Aromatica.

2. Menthol.

Menthol.

$$C_{10}H_{20}O.$$

Botanical and Geographical Sources.—As already noticed,
menthol is stated in the pharmacopœia to be obtained not
only from the oil of peppermint, but also from the oils
derived from varieties of *Mentha arvensis*, DC. Menthol
is largely imported from China and Japan under the names
of *Chinese* and *Japanese Oil of Peppermint* ; the former,
according to Holmes, being obtained from *M. arvensis*,
var. *piperascens,* and the latter from var. *glabrata*, but the

botanical sources can scarcely be said to be absolutely ascertained. The menthol from *Mentha piperita* is sometimes
distinguished as Pip-Menthol, and is now largely prepared
in the United States of America.

General Characters and Tests.—These are given in the
British Pharmacopœia as follows :—In colourless acicular
crystals, usually more or less moist from adhering oil; or in
fused crystalline masses. Its melting-point should not exceed 110° F. (43·3° C.). The hardest masses do not melt
below 108° F. (42·2° C.). It has the odour and flavour of
peppermint, producing warmth on the tongue, or, if air is
inhaled, a sensation of coolness. It is sparingly soluble in
water, and readily soluble in rectified spirit, the solutions
having a neutral reaction. Boiled with sulphuric acid diluted
with half its volume of water, menthol acquires an indigo-
blue or ultramarine colour, the acid becoming brown. It
should be entirely dissipated by the heat of a water-bath.

Medicinal Properties.—Given internally in doses of from
½ to 2 grains, it is a diffusible stimulant. Topically applied
it is largely employed as a remedy in neuralgia, etc. It has
also powerful antiseptic properties.

Dose.—½ to 2 grains.

4. THYMUS VULGARIS.

Garden Thyme. Thyme.

(Bentley and Trimen's ' Medicinal Plants,' vol. iii. plate 205.)

Habitat.—It grows abundantly in Portugal, Spain,
Southern France, Italy, and Greece. It is cultivated in
English gardens as a sweet herb.

Official Product and Name.—THYMOL :—a stearoptene
obtained from the volatile oils of Thymus vulgaris, *Linn.*,
Monarda punctata, *Linn.*, and Carum Ajowan, *Benth. &
Hook.* (Ptychotis Ajowan, *DC.*), by saponifying with caustic

soda and treating the separated soap with hydrochloric acid, or from a distilled fraction of the oil by exposure at a low temperature. It may be purified by recrystallisation from alcohol.

Thymol.

Thymol.

$$C_{10}H_{13}HO.$$

General Characters and Tests.—These are given as follows in the British Pharmacopœia :—Large oblique prismatic crystals having the odour of thyme and a pungent aromatic flavour. They sink in cold water, but on heating the mixture to a temperature of $110°$ to $125°$ F. ($43·3°$ to $52·7°$ C.) they melt and rise to the surface. Slightly soluble in cold water, freely soluble in alcohol, ether, and solutions of alkalies. The crystals volatilise completely at the temperature of a water bath. A solution of thymol in half its bulk of glacial acetic acid, warmed with an equal volume of sulphuric acid, assumes a reddish-violet colour.

Medicinal Properties.—Powerfully antiseptic like carbolic acid.

Dose.—$\frac{1}{2}$ to 2 grains.

5. ROSMARINUS OFFICINALIS.

Rosemary.

(Bentley and Trimen's ' Medicinal Plants,' vol. iii. plate 207.)

Habitat.—Common in the Mediterranean district from Spain to Asia Minor, being abundant in Northern Africa, and also reaching Madeira and the Canary Islands.

Official Product and Name.—OLEUM ROSMARINI :—the oil distilled from the flowering tops.

Oleum Rosmarini.

Oil of Rosemary.

Production and Commerce.—The oil distilled from the flowering tops, as directed in the British Pharmacopœia, is a superior oil to that obtained from the stem and leaves ; but as a general rule very little oil is derived from the former, nearly all the commercial oil being distilled from the stem and leaves of the wild plant, and before it is in flower. Oil of rosemary is principally imported from the South of France and the contiguous Italian coast; but some also comes from Dalmatia by way of Trieste.

General Characters and Composition.—Colourless or pale yellow, with the peculiar odour of rosemary, and a warm aromatic taste. It is neutral, feebly dextrogyre, and freely soluble in alcohol ; its specific gravity has been variously given from 0·896 to 0·908. It is said to be composed of a hydrocarbon, $C_{10}H_{16}$, which constitutes about four-fifths of the whole ; and of an oxygenised oil, and a stearoptene.

Medicinal Properties.—Powerfully stimulant and carminative when administered internally, and largely used as a local stimulant and adjunct to external applications. *Dose.*— 1 to 4 minims.

Official Preparations.

Linimentum Saponis. | Spiritus Rosmarini.
Tinctura Lavandulæ Composita.

Linimentum Saponis is also used in the preparation of Linimentum Opii ; and Tinctura Lavandulæ Composita in that of Liquor Arsenicalis.

SUB-CLASS III. MONOCHLAMYDEÆ OR INCOMPLETÆ.

SERIES I.—*SUPERÆ.*

ORDER I.—POLYGONACEÆ.

1. RHEUM PALMATUM, *Linn.*

2. RHEUM OFFICINALE, *Baillon.*

(Bentley and Trimen's ' Medicinal Plants,' vol. iii. plates 213, 214.)

Habitat.—1. RHEUM PALMATUM.—This plant is a native of the Tangut district of Kansu, the extreme north-western province of China, extending over a large tract of country round the Koko-nor Lake. 2. RHEUM OFFICINALE.—North-eastern part of Thibet near the Chinese frontier, and probably extending into China itself.

Official Part and Name.—RHEI RADIX :—the root, more or less deprived of its bark, sliced and dried, of the above species of Rheum, *Linn.*, and probably other species. Collected and prepared in China and Thibet.

Rhei Radix.

Rhubarb Root.

Production, Preparation, and Commerce.—Rhubarb is produced over a vast area in the Chinese Empire, thus in the four northern provinces of Chihli, Shansi, Shensi, and Honan of China proper ; in Kansu in the north-west, and extending into Thibet ; in the province of Tsing-hai ; and in the western province of Szechuen.

The best rhubarb root is obtained from wild plants, but some is derived from those which are cultivated. But little is known of the preparation of rhubarb root for the market, but in Tangut and other districts where the best rhubarb appears to be obtained, the roots are dug up in September and October from plants from eight to ten years old ; the lateral offsets are then cut off, cleaned, the bark more or less removed, and the roots subsequently cut into pieces ; these are then threaded on strings and suspended in the shade to dry, generally under the roof of a house where the air circulates freely. In other cases they are dried in the sun, but such roots are not uniformly of a firm character throughout. In other districts the pieces are first partially dried on stone tables heated beneath by a fire, and then strung on cords and suspended in the air until the desiccation is complete.

The rhubarb from the districts in which it is thus collected and prepared is principally forwarded to Hankow on the Upper Yangtsze or Blue River, where it is purchased for the European market, and from thence it is sent down to Shanghai and exported to Europe and America. Rhubarb is also exported, although in much smaller quantities, from Tientsin, and in some degree also from Canton, Amoy, and Foochow. Rhubarb root which thus reaches Europe from these different ports presents such varying characters that it seems clear that it is produced from two or more species, probably, therefore, as mentioned in the British Pharmacopœia, not only from *Rheum palmatum* and *Rheum officinale*, but also from other species of *Rheum*.

General Characters.—China Rhubarb, as it is now commonly termed, varies much in form, being somewhat cylindrical, barrel-shaped, conical, plano-convex, or in pieces of irregular form. These varying formed pieces are not, as a rule, found in the same package, but are sorted and distinguished commercially as *flats* and *rounds*. They also vary much in size, but ordinarily from three to four inches long, and from two to three inches broad, but both smaller and larger pieces may be found. The outer surface is more or less covered with a bright yellowish-brown powder ; rounded or marked with flat surfaces and angles ; smooth or more or less shrivelled and wrinkled ; and frequently presenting portions of dark-coloured adherent bark. Beneath the powder the surface is marked with a network of reddish-brown or dark rusty-brown lines, intermixed in a yellowish-brown or yellowish-white substance, and presenting frequently small scattered star-like spots. In many cases the pieces are bored with a hole, which contains the remains of the cord used to suspend them to dry, or in some cases the cord has been removed. When broken transversely the fractured surface is uneven, hard, compact, and presents the same reticulated-marbled appearance as the outer surface, and some pieces exhibit near the cambial line an internal ring of star-like spots, which indicate the points from whence the rootlets arise. The odour is peculiar and somewhat aromatic; and the taste unpleasant, bitter, and feebly astringent, and when chewed it feels gritty between the teeth from the presence of calcium oxalate crystals (*raphides*).

Principal Constituents.—The more important constituents are *chrysophan, emodin, rheo-tannic acid, calcium oxalate,* and three resinous bodies named *aporetin, phæoretin,* and *erythroretin.* The colour is chiefly due to chrysophan, and the same constituent has also been regarded, until lately, as its chief purgative principle, but more recent investigations appear to show that the real active cathartic principle is a substance analogous to the *cathartic acid* of senna. Buchheim therefore concludes that an aqueous extract of rhubarb best represents the activity of the drug; and in this opinion most practical observers are now agreed. The astringent properties of rhubarb depend upon rheo-tannic acid.

Medicinal Properties.—In small doses rhubarb root possesses tonic and slightly astringent properties; and in large doses it acts as a purgative.

Official Preparations.

Extractum Rhei.	Pulvis Rhei Composita.
Infusum Rhei.	Syrupus Rhei.
Pilula Rhei Composita.	Tinctura Rhei.

Vinum Rhei.

OTHER KINDS OF RHUBARB.

As a substitute for the official China Rhubarb, various species of *Rheum* have from time to time been cultivated in Europe for their roots ; but the only kind of special interest to us is that which is commonly known as English Rhubarb, of which from eight to ten tons are prepared annually by Mr. Usher, near Banbury. Until the last few years this was entirely derived from *R. Rhaponticum*, Linn., but now

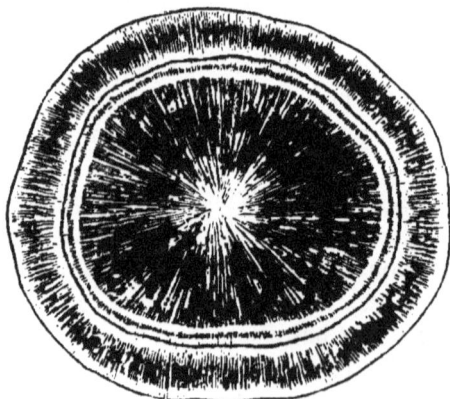

Fig. 38. - Transverse section of a dried cylindrical piece of the root of *Rheum Rhaponticum.*

a small portion is obtained from *R. officinale*, Baillon. Holmes thinks that the plant cultivated at Banbury, *R. Rhaponticum*, is a hybrid with *R. undulatum*, Linn. ; the two species are, however, very closely allied, and should perhaps be united. (Bentley and Trimen's ' Medicinal Plants,' vol. iii. plate 215.)

When well prepared, the pieces of English rhubarb derived from *R. Rhaponticum* are of good colour, and vary in form and size like those of China Rhubarb ; but they may be distinguished from it by being light in weight, more or less spongy in the middle, pasty under the pestle, of a more pinkish hue, and by their internal appearance presenting a less reticulated-marbled appearance, the lines at the ends of the cylindrical pieces being more or less radiate (*fig.* 38), and on the flattened pieces somewhat parallel on the external surface. The star-like spots are also commonly more isolated and irregularly arranged, the taste mucilaginous, more astringent, and usually less gritty when chewed, and the odour feebler and less pleasant.

Order 2.—MYRISTICACEÆ.

MYRISTICA FRAGRANS, *Houttuyn.*

Nutmeg.

Synonym.—Myristica officinalis, *Linn.*

(Bentley and Trimen's ' Medicinal Plants,' vol. iii. plate 218.)

Habitat.—This tree is a native of the Moluccas and other Indian islands—Amboyna, Bouro, New Guinea, etc. It is cultivated in the Banda Islands and the Philippines, and at Bencoolen, and formerly largely at Penang and Singapore, but not now to anything like the same extent; also in Mauritius, the West Indies, and South America.

Official Parts or Products and Names.—1. Myristica :— the dried seed, divested of its hard coat or shell. 2. Oleum Myristicæ :—the oil distilled in Britain from nutmeg. 3. Oleum Myristicæ Expressum :—a concrete oil obtained by means of expression and heat from nutmeg.

1. Myristica.

Nutmeg.

Collection, Preparation, and Commerce.—The fruit is gathered when ripe, and the pericarp removed from the solitary seed it contains, which is covered with an aril; after which the aril, constituting the *mace* of commerce, is carefully stripped from the other parts of the seed (*nut*), and

T

these two parts are then prepared separately for the market. Mace is *not official*, but will be briefly described hereafter ; we now refer to the preparation of nutmegs only. These are prepared as follows :—the nuts are exposed on hurdle-like frames and smoke-dried for about two months at a temperature not exceeding 140° Fahr., being turned every two or three days, and when properly dried the inner portion of the seed, which forms the nutmeg of commerce, separates from its hard coat or shell and rattles when shaken. The shells are then cracked with a wooden mallet, and the nutmegs removed, rubbed over with lime, and packed in tight casks, the insides of which have been smoked, and finally covered with a coating of lime. This is the mode adopted generally (although modified in slight particulars in some districts) by the Dutch in the Banda Islands and Amboyna, and such are called *limed nutmegs.* At Penang and Singapore, however, the nutmegs, after being dried as above, are not coated with lime, and are hence termed *unlimed nutmegs.* These are most esteemed in this country, and constitute the official nutmeg.

Several kinds of nutmegs are known in commerce, the best being *Penang, Singapore,* and *Dutch* or *Batavian* ; the latter being produced in by far the largest quantities.

General Characters.—Nutmegs are oval or roundish, varying somewhat in size, but rarely exceeding one inch in length and four-fifths of an inch in diameter, and in weight averaging about a quarter of an ounce. In colour the unlimed or official nutmegs are uniformly greyish-brown externally ; but the limed nutmegs are brownish in the projecting parts and whitish in the furrows. Nutmegs are smooth to the touch and marked on the outside with a network of furrows ; internally they are greyish-red, marbled with numerous darker reddish-brown lines from the projecting inwards of the inner coat into the albumen. Their odour is strong and pleasantly aromatic ; and their taste agreeably aromatic, warm, and bitterish.

Besides the *true* or *official nutmeg*, as just described, a very inferior nutmeg is sometimes seen in commerce which is the produce of *Myristica fatua*, Houtt. It is readily distinguished by its greater length, and is hence known as the *long nutmeg*, the true nutmeg being characterised as the *round* or *official nutmeg*.

Principal Constituents.—Nutmegs contain about 26 per cent. of fatty substances, the more important of which is *myristin* (*see* Oleum Myristicæ Expressum), and commonly from 2 to 3 per cent. of the official *volatile oil*, but in some cases as much as 8 per cent. has been obtained.

2. Oleum Myristicæ.

Volatile Oil of Nutmeg.

General Characters and Composition.—It is to this oil that the odour and taste of the official nutmeg are essentially due. The oil is colourless or straw-yellow, soluble in both alcohol and ether, dextrogyre, and with a specific gravity which has been variously estimated from 0·920 to 0·948. It is composed of an oxygenated oil, termed *myristicol*, $C_{10}H_{14}O$, and of the hydrocarbon, *myristicene*, $C_{10}H_{16}$.

3. Oleum Myristicæ Expressum.

Expressed Oil of Nutmeg.

Synonym.—Myristicæ Adeps.

Preparation, General Characters, and Composition.—This is obtained by reducing nutmegs to a coarse powder, and then submitting this in a bag to the vapour of hot water, and subsequently pressing between heated plates.

This oil, which is commonly termed *Butter of Nutmeg*, is chiefly derived from Singapore. It is usually found in oblong cakes about ten inches long by two and a half inches square, enveloped in palm leaves. It has an orange-brown or orange-yellow colour, a firm consistence, a more or less

T 2

mottled appearance, the fragrant odour of nutmegs, and
an aromatic fatty taste. It is soluble in about four parts
of warm alcohol, and in two of warm ether. It consists
chiefly of a fat termed *myristin*, and about 6 per cent. of
the volatile oil of nutmeg.

Medicinal Properties.—Nutmegs are aromatic, carminative, and stimulant, and very feebly narcotic. The nutmeg
is largely used as an adjunct to other medicines. The
volatile oil is also stimulant and carminative ; and the
expressed oil is a useful stimulant application.

Official Preparations.

1. Of MYRISTICA :—

Oleum Myristicæ.
Oleum Myristicæ Expres-
sum.
Pulvis Catechu Compositus.
Pulvis Cretæ Aromaticus.

Spiritus Armoraciæ Compo-
situs.
Tinctura Lavandulæ Com-
positus, which is also
used in the preparation
of Liquor Arsenicalis.

2. Of OLEUM MYRISTICÆ :—

Pilula Aloes Socotrinæ.
Spiritus Myristicæ.

Spiritus Ammoniæ Aroma-
ticus.

Spiritus Myristicæ is also used in the preparation of
Mistura Ferri Composita ; and Spiritus Ammoniæ Aro-
maticus for preparing Tinctura Guaiaci Ammoniata, and
Tinctura Valerianæ Ammoniata.

3. Of OLEUM MYRISTICÆ EXPRESSUM :—

Emplastrum Calefaciens. | Emplastrum Picis.

MACIS.

Mace.

(*Not Official.*)

Nature, Preparation, and Commerce.—As already noticed,
mace is the aril of the seed. After its separation from the
nut (*see* page 273) it is dried in the sun, or, in wet weather,

by artificial heat, and is then usually preserved by sprinkling it with salt water. It is commonly derived from the Banda Islands.

General Characters and Composition.—In pieces called *blades*, which are generally orange-yellow, or rarely red, flattish, smooth, irregularly slit, usually flexible, somewhat translucent, and when pressed exuding oil with the odour and taste of nutmegs.

Its essential constituent is from 6 to 9 per cent. of a *volatile oil* resembling that of the volatile oil of nutmeg.

Medicinal Properties.—Same as the nutmeg; but it is principally used as a condiment.

Order 3.—LAURACEÆ.

1. NECTANDRA RODIÆI, *Schomb.*
Bibiru. Greenheart.

(Bentley and Trimen's ' Medicinal Plants,' vol. iii. plate 219.)

Habitat.—It is found only in British Guiana.

Official Parts or Products and Names.—1. NECTANDRÆ CORTEX:—the dried bark. 2. BEBERINÆ SULPHAS:—a salt prepared from Nectandra or Bebeeru bark.

1. Nectandræ Cortex.
Bebeeru Bark.

Commerce.—It is imported from British Guiana, but its supply is very irregular.

General Characters.—In flattish heavy pieces, which are commonly from about one to two feet long, two to six inches broad, and a quarter of an inch or more thick. Its external surface is of a greyish-brown colour, and marked with numerous longitudinal depressions (*digital furrows*) when the suberous layer has been removed; internally it is dark cinnamon-brown, and presents evident longitudinal striæ. It is very hard and brittle, and when broken the fractured surface has a coarse-grained appearance, except internally, where it is somewhat fibrous. It has no marked odour, but a strong bitter astringent taste.

Principal Constituents.—The alkaloids *beberine, nectandrine,* and probably others. Beberine is not, however, a peculiar alkaloid, but, as shown by Walz and Flückiger, it is analogous to *buxine* from *Buxus sempervirens,* and *pelosine* from *Chondrodendron tomentosum.* Sulphate of beberine being official is described below.

Official Preparation.—Beberinæ Sulphas. *Dose.*—1 to 10 grains.

2. Beberinæ Sulphas.

Sulphate of Beberine.

Synonym.—Beberiæ Sulphas.

Preparation.—This salt, as already stated, is prepared from nectandra or bebeeru bark, which is official for that purpose. It is probably a mixture of sulphates of beberine, nectandrine, and other alkaloids ; and a process for its preparation is given in the pharmacopœia.

General Characters and Tests. — In dark-brown thin translucent scales, yellow when in powder, with a strong bitter taste, soluble in water, yielding a clear brown solution, and in alcohol. Its watery solution gives a white precipitate with chloride of barium; and with caustic soda a yellowish-white precipitate, which is dissolved by agitating the mixture with twice its volume of ether. The ethereal solution, separated by a pipette and evaporated, leaves a yellow translucent residue, entirely soluble in dilute acids. Ignited with free access of air it burns without residue.

Medicinal Properties.—Bebeeru bark has astringent and tonic properties ; but it is scarcely ever used. Sulphate of beberine is regarded as tonic and antiperiodic, but it is little employed.

2. SASSAFRAS OFFICINALE, *Nees.*
Sassafras.

(Bentley and Trimen's ' Medicinal Plants,' vol. iii. plate 220.)

Habitat.—The tree is common in Canada and the United States of America, as far south as Florida and Missouri.

Official Part and Name.—SASSAFRAS RADIX :—the dried root, reduced to chips or shavings.

Sassafras Radix.
Sassafras Root.

Production and Commerce.—Sassafras root is obtained from the United States of America, the principal mart for it being Baltimore. As imported it is in branched logs or billets, more or less covered with bark, and frequently having attached a portion of the lower part of the stem. The latter should be removed as it is almost inert, and then the root cut into chips or shavings for use in pharmacy.

General Characters.—The wood of the root is soft, light in weight, dull greyish-red or greyish-yellow, and coarse-grained. The bark with which it is more or less covered is rough, and greyish-brown or rusty-brown externally, and internally smooth, glistening, and dark rusty-brown. Both the wood and the bark have an agreeable aromatic odour, and a peculiar aromatic somewhat astringent taste ; the odour and taste of the wood being more feeble than the bark. In the pharmacies it is commonly found in the form of chips or shavings.

Principal Constituents.—Its essential constituent is a *volatile oil*, the amount obtainable being from 1 to 2 per cent., depending in a great measure upon the proportion of bark—the latter containing most oil. The oil is yellowish or brownish when fresh, but becomes reddish by age ; it has a pungent aromatic taste, and a similar odour to the root. It is feebly dextrogyrate, and has a high specific gravity, being about 1·090. *Tannic acid* is also a constituent of sassafras root.

Medicinal Properties.—Sassafras root is aromatic, stimulant, and diaphoretic. The oil has similar properties but is not official in Britain.

Official Preparation.—Decoctum Sarsæ Compositum.

---

3. LAURUS NOBILIS, *Linn.*

Sweet Bay. True Laurel.

(Bentley and Trimen's ' Medicinal Plants,' vol. iii. plate 221.)

Habitat.—It is found wild in Asia Minor and Syria. It grows also in the countries surrounding the Mediterranean, and is abundant in Greece, Italy, and Southern France ; it is, however, generally considered to have been introduced from the East.

Part Used and Name.—LAURI FRUCTUS :—the fresh and dried fruit (*drupe*). These fruits are improperly termed *bay berries.*

(*Not Official.*)

Lauri Fructus.

Bay Berries.

Commerce.—These fruits are commonly derived from Southern Europe.

General Characters.—When dried the fruits are about one-third of an inch long, and oval in form ; they consist of a thin blackish- or dark purplish-brown brittle wrinkled pericarp, which is easily removed ; containing a solitary loose firm seed, of the same form as the fruit, and easily separated into two equal lobes. The pericarp has no marked odour or taste ; but the seed has an agreeable aromatic odour, and a bitter aromatic taste. For the mode of distinguishing these fruits from 'Cocculus Indicus,' with which they have been confounded, see 'Anamirta Cocculus.'

Principal Constituents.—Their properties depend upon the presence of a fragrant *volatile oil* and a *concrete fixed oil.* The latter is imported from the South of Europe, and is commonly known in commerce as *Oil of Bays.* It is usually obtained by submitting the fresh fruits, previously steeped in hot water, to pressure. It is a mixture of fatty bodies and volatile oil, like the expressed oil of nutmeg already described.

Medicinal Properties.—The fruits are aromatic, stomachic, and stimulant, but are not now used in this country. The oil of bays is used externally as a stimulant application, but only in veterinary practice. The volatile oil has similar properties.

---

4. CINNAMOMUM CAMPHORA, *Nees & Eberm.*

Camphor. Laurel Camphor.

Synonym.—CAMPHORA OFFICINARUM, *C. G. Nees.*

(Bentley and Trimen's 'Medicinal Plants,' vol. iii. plate 222.)

Habitat.—Common in Japan and Formosa, and also abundant in Central China.

Official Product and Name.—CAMPHORA :—a stearoptene obtained from the wood. Imported in the crude state, and purified by sublimation.

Camphora.
Camphor.

I. CRUDE CAMPHOR.—*Production and Commerce.*—Camphor in a crude state is derived from the Island of Formosa and from Japan ; the former is characterised as *China* or *Formosa Camphor*, and the latter as *Japan* or *Dutch Cam-*

phor. Both kinds are obtained by submitting chips of camphor wood to heat, either by boiling them in water or by exposing them to the vapour of boiling water, and then collecting the camphor which volatilises with the steam and is subsequently condensed in suitable receptacles. Formosa camphor is then stored in vats or barrels, when a brownish-yellow volatile oil drains out. This is the *oil of camphor*, now frequently used in this country and elsewhere as a stimulant application in rheumatism, etc. Before exportation water is commonly poured upon the camphor, with a view of lessening its evaporation.

General Characters.—Crude camphor from *Formosa* is in moist friable masses, of irregular forms and sizes, and consisting of greyish-white or light-brown, somewhat crystalline grains. *Japanese camphor* is in larger grains, which have commonly a pink hue, and are cleaner and drier than Formosa camphor and of higher commercial value.

2. REFINED CAMPHOR.—*Process of Refining.*—The crude camphor is commonly refined in this country by mixing it in glass vessels, called *bombaloes* (*fig.* 39), with a little lime, and submitting it carefully to heat; and when sublimed in this way it is obtained in the form of concavo-convex cakes or bowls, each of which is about ten inches in diameter, and from nine to twelve pounds in weight. Other modes of sublimation are also adopted, and different formed vessels are likewise employed for that purpose, but no camphor unless refined *is official.*

FIG. 39. Bombalo in which crude Camphor is commonly refined.

General Characters. —Refined camphor is in solid colourless translucent crystalline masses or pieces, of varying forms, which present numerous fissures when of any size. The pieces are somewhat tough, but readily powdered if moistened with rectified spirit, ether, or chloroform. Camphor floats on water, burns readily with a bright smoky flame, volatilises somewhat rapidly even at ordinary temperatures, and sublimes entirely when heated; it is very slightly soluble in water, but readily soluble in rectified spirit, ether, or chloroform. It has a powerful penetrating odour; and a pungent somewhat bitter taste, followed by a sensation of cold. Its specific gravity from 50° to 54° F. is about 0·992, but between 32° and 42° F. nearly that of water. When melted or in a concentrated solution it is dextrogyrate.

Composition.—Camphor has a composition of $C_{10}H_{16}O$; it belongs to the class of *stearoptenes* or *solid volatile oils*.

Medicinal Properties.—In proper doses, camphor is stimulant, sedative, diaphoretic, and antispasmodic; and in large doses it is an acro-narcotic poison. Camphor has likewise antiseptic properties, and is sometimes regarded as an anaphrodisiac. Locally applied it is stimulant and anodyne. *Dose.*—1 to 10 grains.

Official Preparations containing Camphor.

Aquæ Camphoræ.

Injectio Apomorphinæ Hypodermica.

Injectio Ergotini Hypodermica.

Linimentum Aconiti.

Linimentum Belladonnæ.

Linimentum Camphoræ.

Linimentum Camphoræ Compositum.

Linimentum Chloroformi.

Linimentum Hydrargyri.

Linimentum Opii.

Linimentum Saponis.

Linimentum Sinapis Compositum.

Linimentum Terebinthinæ.

Linimentum Terebinthinæ Aceticum.

Spiritus Camphoræ.

Tinctura Camphoræ Composita.

Unguentum Hydrargyri Compositum.

5. CINNAMOMUM ZEYLANICUM, *Breyn.*

Cinnamon.

(Bentley and Trimen's ' Medicinal Plants,' vol. iii. plate 224.)

Habitat.—It is a native of Ceylon, where it is also extensively cultivated. It has been introduced into India, Java, Brazil, West Indies, etc.

Official Parts or Products and Names.—1. CINNAMOMI CORTEX :—the dried inner bark of shoots from the truncated stocks or stools of the cultivated cinnamon tree. Imported from Ceylon, and distinguished in commerce as Ceylon Cinnamon. 2. OLEUM CINNAMOMI :—the oil distilled from cinnamon bark.

1. Cinnamomi Cortex.

Cinnamon Bark.

Production and Commerce.—The official cinnamon bark is directed to be obtained from trees cultivated in Ceylon, and is distinguished in commerce as *Ceylon Cinnamon.* The bark is prepared as follows : The shoots are removed, then divested of their bark, and the latter, after being left for about twenty-four hours, is carefully scraped so as to remove its two outer layers, after which the smaller quills are introduced within the larger ones in such a way as to form congeries of quills (compound quills) of about forty inches in length. The bark is then dried by exposure first in the shade, and afterwards in the sun, and finally made up into bundles for exportation. The official bark consists therefore essentially of the inner bark or liber.

Besides the official *Ceylon Cinnamon* other varieties are distinguished in commerce as *Malabar* or *Tinnivelly*, *Telli-cherry*, and *Java*, all of which are of lower commercial value, and of inferior quality. These kinds are imported in far less quantities.

General Characters and Test.—In closely rolled quills, each being about three-eighths of an inch in diameter, and containing several smaller quills. The bark itself is thin, brittle, with a splintery fracture, moderately pliable, of a dull light yellowish-brown colour externally, and marked at intervals by little holes or scars, indicating the points from whence the leaves have been removed, and faint shining wavy lines ; internally it is darker brown. The odour is very fragrant ; and the taste warm, sweet, and aromatic. A decoction, when cool, is not coloured by iodine.

Inferior kinds of cinnamon bark are thicker, darker coloured, and have a more pungent and somewhat bitter taste.

Principal Constituents.—The only important constituent is the *volatile oil*, which being official is described below.

2. Oleum Cinnamomi.

Oil of Cinnamon.

Commerce. — Oil of Cinnamon is imported from Ceylon.

General Characters. — Yellowish when recent, but gradually becoming cherry-red, and with the odour and taste of cinnamon bark. Its specific gravity is about 1·035. It sinks in water.

Medicinal Properties.—Cinnamon bark is aromatic, carminative, stimulant, and somewhat astringent. The oil of cinnamon has the cordial and stimulant properties of the bark without its astringency. Both cinnamon bark and oil of cinnamon are commonly used as adjuncts to other medicines as flavouring agents or as correctives.

Official Preparations.

1. Of CINNAMOMI CORTEX :—

Aquæ Cinnamomi.
Decoctum Hæmatoxyli.
Infusum Catechu.
Oleum Cinnamomi.
Pulvis Catechu Compositus.
Pulvis Cinnamomi Compo-
    situs.
Pulvis Cretæ Aromaticus.

Pulvis Kino Compositus.
Tinctura Cardamomi Com-
    posita.
Tinctura Catechu.
Tinctura Cinnamomi.
Tinctura Lavandulæ Compo-
    sita.
Vinum Opii.

Aqua Cinnamoni is also used in the preparation of Mistura Cretæ, Mistura Guaiaci, and Mistura Spiritus Vini Gallici. Pulvis Cinnamomi Compositus is contained in Pilula Aloes et Ferri, and Pilula Cambogiæ Composita. Tinctura Cardamomi Composita is used in four other official preparations ; and Tinctura Lavandulæ Composita in the preparation of Liquor Arsenicalis.

2. Of OLEUM CINNAMOMI :—Spiritus Cinnamomi, which is also used in the preparation of Acidum Sulphuricum Aromaticum; the latter being also a constituent of Infusum Cinchonæ Acidum.

Cassia Lignea.

Cassia Bark.

(Not Official.)

Cassia bark, which is derived from *Cinnamomum Cassia*, Blume (Cinnamomum aromaticum, *Nees*), is sometimes substituted for cinnamon bark, and is official, like it, in the United States Pharmacopœia. There are several varieties of cassia bark, but the best comes from China, and on the Continent is even termed *Chinese Cinnamon*.

It may be readily distinguished from official cinnamon bark by its quills being less closely rolled and single or rarely double, by portions of the outer coat being left on their surface, by its thicker substance, coarser appearance, darker brown and duller colour, and by its less sweet and more pungent and bitter taste, and by its less fragrant odour. Another ready way of distinguishing cassia bark from cinnamon bark is by solution of iodine, for while a decoction of cinnamon bark when cool is not coloured by iodine, a blue colour is at once produced when this is added to a cooled decoction of cassia bark.

ORDER 4.—THYMELACEÆ.

1. DAPHNE MEZEREUM, *Linn.*

Mezereon.

(Bentley and Trimen's ' Medicinal Plants,' vol. iii. plate 225.)

2. DAPHNE LAUREOLA, *Linn.*

Spurge Laurel.

(Bentley and Trimen's ' Medicinal Plants,' vol. iii. plate 226.)

Habitat.—1. DAPHNE MEZEREUM.—This is a native of hilly woods of almost the whole of Europe, from Italy to the arctic regions, and extends eastwards into Siberia. It is very rare in England, and probably not indigenous. 2. DAPHNE LAUREOLA.—This species is not uncommon in England south of Durham ; but probably not indigenous in Scotland. It is also found throughout Western and Southern Europe ; and in the Azores, Algeria, and Asia Minor.

Official Part and Name.—MEZEREI CORTEX : – the dried bark of the above two plants.

Mezerei Cortex.

Mezereon Bark.

Collection and Commerce.—Mezereon bark is collected in the winter months, dried, and then made up into rolls or bundles of various sizes and lengths. It is principally imported from Germany, although formerly it was obtained in this country from Kent and Hampshire.

General Characters.—In small quills of various lengths ; or, more commonly, in long thin more or less flattened strips or pieces, which are usually folded or rolled into disks. It is covered externally by an olive-brown or somewhat reddish-brown, readily separable corky layer ; and internally it is whitish or yellowish-white, silky or somewhat cottony, and very tough and difficult to tear asunder. The *stem-bark* is commonly known by being in evident quills, by its darker coloured outer surface, which is marked by leaf-scars, and by the green colour of its cellular envelope (*phelloderm*) or part beneath the outer corky layer. Neither the stem-bark nor root-bark, when dried, has any marked odour ; but when chewed their taste is burning and acrid, that of the stem-bark being somewhat less evident.

Principal Constituents.—Its principal ingredient is an *acrid resin* ; it appears also to contain a little *acrid volatile oil*, but this requires further examination.

Medicinal Properties.—Locally applied it is sialagogue, rubefacient, and vesicant ; and when administered internally stimulant, diaphoretic, alterative, and diuretic. In large doses it is an irritant poison.

Official Preparations.

Decoctum Sarsæ Compositum.

Extractum Mezerei Æthereum, which is used in the preparation of Linimentum Sinapis Compositum.

ORDER 5.—MORACEÆ.

1. FICUS CARICA, *Linn.*
Fig.

(Bentley and Trimen's ' Medicinal Plants,' vol. iv. plate 228.)

Habitat.—It is a native of Syria and some adjacent parts of Asia Minor, and apparently extending in a wild state to the north-west confines of India. It is now found in cultivation in all the temperate and warmer countries of both hemispheres.

Official Part and Name.—FICUS :—the dried fruit.

Ficus.
Fig.

Preparation and Commerce.—The figs of commerce are gathered when ripe, dried in ovens, or by exposure to the sun ; and then either packed without further preparation, in which case they are termed *natural figs* ; or rendered pliant by squeezing and kneading, and packed by pressing into drums or boxes, when they are known as *pulled figs.* The best figs come from Smyrna ; but some reach us from Greece, Spain, Portugal, and other countries. The Smyrna figs are pulled figs. In the London shops the finest figs are sometimes distinguished as *Eleme figs.*

General Characters.—The fig consists of the enlarged hollow succulent receptacle, which has a small opening at its apex, and bears upon its inner surface a great number of small seed-like fruits or achenes, the whole together forming the kind of anthocarpous fruit termed a *syconus.* The dried fig is irregular in form, compressed, soft, tough, more or less translucent, brownish or yellowish externally, and covered, at least in cool weather, with a saccharine efflorescence. The odour is fruity and pleasant ; and the taste sweet, peculiar, and most agreeable.

Principal Constituents.—*Grape sugar* is a principal con-

U

stituent, and forms from 60 to 70 per cent. of the dried fruits of commerce.

Medicinal Properties.—Figs are regarded as demulcent and laxative when administered internally ; and emollient when locally applied.

Official Preparation.—Confectio Sennæ.

---

2. MORUS NIGRA, *Linn.*
Mulberry.

(Bentley and Trimen's ' Medicinal Plants,' vol. iv. plate 229.)

Habitat.—It is supposed to be a native of Persia, the Caucasus, and Armenia.; and to have spread westward in early times.

Official Part and Name.—MORI SUCCUS :—the juice of the ripe fruit.

Mori Succus.
Mulberry Juice.

General Characters of the Fruit and Juice.—The fruits are an illustration of the kind of anthocarpous fruit termed a *sorosis.* When ripe they have a dark purple or almost black colour, and a refreshing acidulous saccharine taste. These fruits, commonly known as *mulberries*, are very juicy, and when this juice is extracted it has a dark violet or purple colour, a faint odour, and a similar taste to the fruit. Its specific gravity is about 1·060.

Principal Constituents.—Mulberry juice is rich in *grape sugar*, which is its principal constituent ; but it contains also some free acid, commonly supposed to be *malic acid*, but probably, in part at least, *tartaric.*

Medicinal Properties.—Mulberry juice is regarded as refrigerant and slightly laxative. It is chiefly employed as a colouring and flavouring agent, and as a laxative for infants.

Official Preparation.—Syrupus Mori.

ORDER 6.—CANNABINACEÆ.

1. HUMULUS LUPULUS, *Linn.*

Hop.

(Bentley and Trimen's ' Medicinal Plants,' vol. iv. plate 230.)

Habitat.—It is found in a wild state throughout Europe, except the extreme north, and extends eastward to the Caucasus, and through Central Asia to the Altai Mountains. It is a common but doubtfully indigenous plant in the Northern and Western United States of America, and has been introduced into Brazil and Australia.

Official Parts and Names.—1. LUPULUS :—the dried strobiles. 2. LUPULINUM :—a glandular powder obtained from the dried strobiles.

1. Lupulus.

Hop.

Synonym.—Humulus.

Collection and Preparation.—The anthocarpous or collective fruits, termed strobiles, are directed to be obtained from plants cultivated in England. For this purpose the gathering or picking takes place in September, chiefly in Kent and the adjacent parts of Sussex and Surrey, but also to some extent in Hampshire, Worcestershire, and Herefordshire ; they are then dried in kilns, and finally packed in hempen sacks, called *bags* or *pockets*, for the market.

General Characters.—The strobiles, as found in commerce, are more or less compressed and broken up. When entire, they are about one inch and a quarter long ; oblong-ovoid or rounded in form, and consisting of a number of thin, greenish-yellow or brownish, membranous, ovate or roundish, imbricated bracts or scales; each of which has at its base, on its inner surface, a small rounded seed-like

U 2

achene, which is sprinkled over with brownish-yellow glands (*lupulin*), the whole being attached to a hairy undulating axis or peduncle. They have an agreeably aromatic odour, and a bitter, aromatic, and feebly astringent taste.

2. Lupulinum.

Lupulin.

Synonym.—Lupulinic Glands.

Collection and Preservation.—These glands, as we have seen, are more especially found covering the achenes placed at the base of the bracts or scales on their inner surface, but to a slight extent also over other parts. The dried strobiles yield from 8 to 16 per cent. of these glands. They are readily separated from the scales by rubbing and shaking, and then separating from other parts by passing them through a sieve. They should be subsequently washed by decantation to free them from any earthy matters which always adhere to them to some extent; and then carefully dried and preserved in well-stoppered bottles.

General Characters and Tests.—When viewed in substance by the naked eye, the glands form a bright brownish-yellow granular powder. Lupulin burns readily when thrown into the air and ignited ; and has the agreeable aromatic odour and bitter aromatic taste of hop. When examined by the microscope, it is seen to consist of minute, somewhat globular - top-shaped, reticulated, translucent, shining glands. Each grain is attached by a short pedicel to the scales (*fig.* 40), but this stalk is scarcely perceptible in the dried glands of commerce. When fresh the gland contains a dark brownish-yellow liquid, but when dried this liquid is contracted so as to form a mass in its centre. On incineration lupulin should not yield more than about 10 per cent.

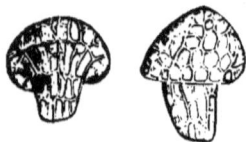

FIG. 40. Fresh Lupulinic Glands. (After Maisch.)

of ash. Not more than about 30 or 40 per cent. should be insoluble in ether.

Principal Constituents.—The principal constituents of hop are *volatile oil, bitter principle, wax,* and *resin* ; these substances exist more especially in the lupulinic glands. The scales only contain a trace of the bitter principle, and but a very small proportion of the volatile oil ; but one of their constituents, which is not present in the glands, is a kind of *tannic acid,* termed *humulo-tannic acid.* The aromatic and bitter properties of hop reside therefore essentially in the glands ; and their astringency in the scales or bracts. An amorphous substance of a dark red colour, termed *phlobaphene,* is also found in the scales, but not in the glands.

The *volatile oil* exists in the glands in the proportion of about 2·5 per cent., and in the hop generally to about 0·9 per cent. It yields by exposure to air valerianic acid. The *bitter principle* has been variously termed *lupuline, lupulite, bitter acid of hops,* and *lupamaric acid.* It crystallises in large brittle rhombic prisms, and is almost insoluble in water. In the glands it occurs in the proportion of from 8·5 to about 12 per cent. The proportion of *humulo-tannic acid* in the scales is from about 3 to more than 4 per cent.

Medicinal Properties.—Hops possess tonic, stomachic, and slightly soporific properties, which essentially reside in their lupulinic glands; but they are also somewhat astringent. Hops added to beer and ale check acetous fermentation.

Official Preparations.

1. Of Lupulus :—

     Extractum Lupuli | Infusum Lupuli
                   Tinctura Lupuli.

2. Of Lupulinum :—

     There are no preparations. *Dose.*—2 to 5 grains.

2. CANNABIS SATIVA, *Linn.*
Hemp.   Indian Hemp.

(Bentley and Trimen's 'Medicinal Plants,' vol. iv. plate 231.)

Habitat.—It is a native of Central and Western Asia. It was early introduced into Italy, and has gradually spread into all temperate and warm countries of both the Old and the New Worlds.

Official Part and Name.—CANNABIS INDICA :—the dried flowering or fruiting tops of the female plants, grown in India, and from which the resin has not been removed.

Cannabis Indica.
Indian Hemp.

Production and Commerce.—There are several forms and preparations of Indian Hemp, but the *only one which is official* is that known in India under the Hindustani name of *Gunjah* or *Ganja*, and by the London drug brokers as *Guaza*. Bengal Ganja is said to be very superior to that produced in other parts of India.

General Characters.—In small more or less aggregated masses, from about one and a half to two and a half inches in length, and consisting of the tops of one or more alternate branches bearing the remains of the flowers and smaller leaves with a few ripe fruits, and the whole pressed together by adhesive resinous matter ; or it is composed of straight stiff woody stems several inches long, surrounded by the branched flower-stalks.   It is rough to the touch, very brittle, of a dusky-green or brownish-green colour, with only a very feeble acrid taste ; but having a faint, peculiar, narcotic, not unpleasant odour.

Principal Constituents.—A little *volatile oil* ; and a *brown amorphous resin*, to which its narcotic properties are said to be due, but this has not been definitely determined. The presence of an alkaloid, *tetano-cannabin*, has also been indicated as a constituent of Indian Hemp, but it requires confirmation.

Medicinal Properties.—Anodyne, soporific, antispasmodic, and nervine stimulant.

Official Preparations. — Extractum Cannabis Indicæ. *Dose.*—¼ to 1 grain. This extract is used in the preparation of Tinctura Cannabis Indicæ, the dose of which is from 5 to 20 minims.

ORDER 7.—ULMACEÆ.

ULMUS CAMPESTRIS, *Linn.*

Elm.

(Bentley and Trimen's ' Medicinal Plants,' vol. iv. plate 232.)

Habitat.—This tree is widely distributed throughout Central, Southern, and Eastern Europe, extending southward to North Africa and Asia Minor, and eastward as far as Northern China and Japan. It is not truly indigenous in this country.

Part Used and Name.—ULMI CORTEX :—the dried inner bark.

(*Not Official.*)

Ulmi Cortex.

Elm Bark.

Collection and Preparation.—It should be collected in the spring from trees indigenous to, and cultivated in, Britain. The bark after its separation from the trees should have its corky and middle layers removed, and the liber which then remains should be quickly dried.

General Characters.—The bark obtained as just described, is in broad flattish pieces of varying lengths and thickness, but rarely more than one eighth of an inch. Its outer surface has a brownish-yellow or rusty-yellow colour,

and it is striated on its inner surface.   It is marked externally
by nearly smooth portions (caused by the knife in removing
the outer layers), separated by slight projecting angles.   Elm
bark is tough and somewhat fibrous ; almost inodorous, and
with a slightly mucilaginous and bitter taste, combined with
astringency.

Principal Constituents.—*Mucilage* and *tannic acid*, to
which it owes its properties.   It also frequently contains *starch*,
owing to its middle layer not having been entirely removed,
but there is no starch in the liber.

Medicinal Properties.—Mild astringent, tonic, and demulcent ; and also in full doses it is said to be diuretic and
diaphoretic.   As a medicine, however, in this country it is
nearly obsolete, and is therefore no longer official.

Preparation.—The best form of administration is the
Decoctum Ulmi of the British Pharmacopœia of 1867.

ORDER 8.—SALICACEÆ.

SALIX ALBA, *Linn.*

White Willow.   Golden Willow.

(Bentley and Trimen's 'Medicinal Plants,' vol. iv. plate 234.)

Official Product and Name.—SALICINUM :—a crystalline
glucoside obtained by treating the bark of Salix alba, *Linn.*,
and other species of Salix, *Linn.*, and the bark of various
species of Populus, *Linn.*, with hot water, removing tannin
and colouring matter from the decoction, evaporating,
purifying, and recrystallising.

Salicinum.

Salicin.

$$C_{13}H_{18}O_7.$$

General Characters and Tests.—These are given in the
British Pharmacopœia as follows :—Colourless shining
crystals with a very bitter taste.   Soluble in about twenty-

eight parts of water or sixty-five parts of spirit at common temperatures ; insoluble in ether. Sulphuric acid colours it red. A small quantity heated with a little red chromate of potassium, a few drops of sulphuric acid, and some water, yields vapours of an oil having the odour of meadow-sweet. The crystals melt when heated, and emit vapours having the odour of meadow-sweet. On ignition in air it leaves no residue.

The *official salicylic acid* was formerly obtained from the willow, and is now described in the British Pharmacopœia as a crystalline acid obtained by the combination of the elements of carbolic acid with those of carbonic acid gas and subsequent purification, or from natural salicylates such as the oils of winter-green (Gaultheria procumbens, *Linn.*) and sweet birch (Betula lenta, *Linn.*).

Medicinal Properties.—*Salicin* in small doses is regarded as a mild tonic and antiperiodic; and in large doses it has an almost specific action in acute rheumatism. *Dose.*—3 to 20 grains.—*Salicylic acid* and its salts, more especially the official salicylate of soda, in doses of from 5 to 30 grains, is now very largely and successfully used in acute rheumatism; it also possesses powerful antiseptic properties.

ORDER 9.—EUPHORBIACEÆ.

1. MALLOTUS PHILIPPINENSIS, *Müll. Arg.*
Kamala.

Synonym.—ROTTLERA TINCTORIA, *Roxb.*

(Bentley and Trimen's ' Medicinal Plants,' vol. iv. plate 236.)

Habitat.—This plant is said to be widely distributed in the East, and is especially abundant in India in the Sub-Himalayan tract.

Official Part and Name.—KAMALA :—a powder which consists of the minute glands and hairs obtained from the surface of the fruits.

Kamala.

Kamala.

Collection and Commerce.—Kamala is imported into this country from India. Kamala is obtained by throwing the fruits into a basket, then by subsequent rolling and rubbing it is separated as a powder, which falls through the basket as through a sieve, and is received below on a cloth spread for that purpose.

General Characters and Test.—Kamala as usually found in commerce consists of small glands, greyish stellate hairs, minute fragments of leaves, remains of the fruits, and earthy impurities. The latter should be separated by sifting, and then kamala is a fine granular mobile powder of a brick-red or madder colour, and nearly inodorous and tasteless. It floats on water, and ignites with a flash when sprinkled over a flame, like lycopodium. It is almost insoluble in boiling water, and entirely so in cold water, but readily soluble in alcohol, ether, or chloroform, and forming deep-red solutions. On ignition in air it should yield 4 or 5, or at most 10, per cent. of ash. When examined by the microscope it is seen to consist of irregular spherical flattened or depressed garnet-red glands with wavy surfaces (*fig.* 41, *a*), mixed with nearly colourless or somewhat greyish thick-walled stellate hairs (*fig.* 41, *b*).

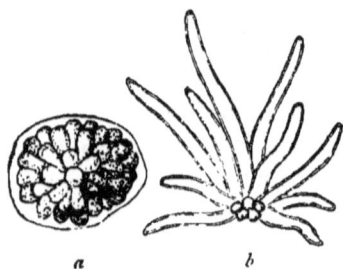

Fig. 41.—Kamala (magnified 190 diameters), showing the glands, *a*, and stellate hairs, *b*. (After Maisch.)

Adulterations and Substitutions.—Kamala is often very largely adulterated with earthy substances, and foreign vegetable matters ; and at times, other drugs under the same name have been substituted for it. The characters given above will at once serve to detect such adulterations and substitutions.

Principal Constituents.—It contains about 80 per cent. of resins, which Leube regards as the active constituents. A crystalline principle was found by Anderson and termed *rottlerin* ; but Leube was unable to isolate this principle. Other chemists have, however, obtained minute crystals from kamala, and hence further experiments are necessary.

Medicinal Properties.—An active purgative, and regarded by many as a valuable anthelmintic against tapeworm. *Dose.* —30 grains to ¼ oz. in powder or electuary.

2. RICINUS COMMUNIS, *Linn.*
Palma-Christi.　Castor Oil.
(Bentley and Trimen's ' Medicinal Plants,' vol. iv. plate 237.)

Habitat.—This plant is supposed to have been originally a native of India, from whence it has spread over all the warmer parts of the world. It is cultivated largely in India, Italy, and other countries, for medicinal purposes.

Official Product and Name.—OLEUM RICINI :—the oil expressed from the seeds.

General Characters of the Seeds.—The seeds are oval, somewhat compressed, convex on one side, and with two flattish surfaces on the other, and marked at one end by a fleshy protuberance termed a caruncule or strophiole, or if this has been broken off, a blackish scar remains. They are from one-third to over half an inch long, and from about a quarter to four-tenths of an inch broad ; they are covered by a smooth shining testa, of a greyish colour, mottled with brownish or blackish bands and spots of various tints and shapes so as to give much variation to the seeds in appearance. The seed-coats are tasteless and inodorous ; but when fresh the nucleus has a bland sweetish taste, followed by a slight acridity.

Principal Constituents.—The essential constituent of the nucleus is the *official fixed oil* described below, of which it yields from 40 to 50 per cent.

Oleum Ricini.

Castor Oil.

Production and Commerce.—All the oil now used medicinally in Britain is obtained by first crushing the seeds, removing the seed-coats, and then submitting the nucleus to pressure. In India, where it is most extensively produced, the oil thus obtained is then heated with water, by which the albuminous matters are separated as a scum, and the oil is finally strained through flannel ; while in the North of Italy, where very good oil is now largely produced, the seeds in the winter are pressed in a warmed room, and in all cases iron plates heated to about 90° F. are placed between the press-bags so as to facilitate the outflow of oil. In expressing castor oil, all processes in which a high temperature is employed are regarded as objectionable in consequence of increasing the acridity of the oil.

Principal Constituents.—Castor oil consists essentially of two *fatty acids* ; one of which is said to be *palmitic acid*, and the other, which is peculiar to it, is known as *ricinolei̇c acid*. These acids are in combination with *glyceryl*. It also contains a very minute proportion of some *acrid principle*, but this has not been isolated. Tuson describes an alkaloid, termed *ricinine*, as a constituent, but its presence has not been confirmed.

General Characters of the Oil.—Viscid, colourless or pale-straw yellow, with scarcely any odour, and a mild taste at first, but subsequently acrid and unpleasant. It has a specific gravity of about 0·96 at 60° F., congeals at about 0° F., and when exposed to the air in thin layers, it slowly dries up to a varnish. It is entirely soluble in one volume of absolute alcohol, and in four volumes of rectified spirit. Inferior oils have a brownish colour, a nauseous odour, and an acrid very unpleasant taste.

Medicinal Properties.—Castor oil is a mild and most efficient non-irritating purgative. *Dose.* 1 to 8 fluid drachms.

Official Preparations.

Collodium Flexile. | Linimentum Sinapis Compositum.
Pilula Hydrargyri Subchloridi Composita.

---

3. CROTON ELUTERIA, *J. J. Bennett.*
Cascarilla.   Sweet Bark.

(Bentley and Trimen's ' Medicinal Plants,' vol. iv. plate 238.)

Habitat.—This plant is found in all the islands of the
Bahama group, and in Cuba.

Official Part and Name.—CASCARILLÆ CORTEX :—the
dried bark.

Cascarillæ Cortex.
Cascarilla Bark.

Commerce.—Cascarilla bark is imported, usually packed
in sacks, from Nassau, in New Providence, one of the
Bahama Islands.

General Characters.—In quills or more or less curved
pieces, which vary from about one to three or more inches
in length, and from one-sixth to half an inch in diameter.
They are covered with a dull-brown readily separable corky
layer of varying degrees of roughness according to its age,
and more or less coated, sometimes entirely, with a silvery-
or greyish-white lichen (*Verrucaria albissima*, Ach.), with
its perithecium in the form of small black dots; the inner
surface is brown and smooth, or but faintly striated. It
has a compact texture, and breaks with a short resinous
fracture of a brown colour. The taste is warm and nauseously bitter ; and the odour agreeable and aromatic, more
especially when burned.

Adulteration.—A spurious cascarilla bark has been noticed
in the London market by Holmes, who suggests that it is
probably derived from *Croton lucidus*, Linn. Like true

cascarilla, which it much resembles, it is more or less covered by a minute lichen ; but its periderm does not readily peel off, and it is more greyish in colour. Its inner surface is pinkish-brown or reddish, and distinctly striated longitudinally; and its taste is astringent, but without bitterness or aroma, and hence readily distinguishable from the true bark.

Principal Constituents.—*Volatile oil* in the proportion of about 1·2 per cent.; *resin*; and a bitter crystalline principle, termed *cascarillin.*

Medicinal Properties.—Stimulant, tonic, and febrifuge. On account of its agreeable odour when burned, it is a useful ingredient in fumigating pastilles.

Official Preparations.

Infusum Cascarillæ. | Tinctura Cascarillæ.

4. CROTON TIGLIUM, *Linn.*

Purging Croton.

(Bentley and Trimen's ' Medicinal Plants,' vol. iv. plate 239.)

Habitat.—It is common both in a wild and cultivated state throughout the Indian peninsula ; it also grows in Ceylon, Borneo, and the Philippines, and has been introduced into Japan and Mauritius.

Official Product and Name.—OLEUM CROTONIS :—the oil expressed in Britain from the seeds.

General Characters of the Seeds.—Croton seeds are oval, somewhat quadrangular, from about one-half to three-fifths of an inch long, dark reddish-cinnamon-brown on the outer surface, and when scraped, black. The testa is brittle, and encloses a pale-coloured delicate silky tegmen or inner coat, and a yellowish oily nucleus. The seeds have no odour; and but little taste at first, but subsequently burning and acrid.

Principal Constituents.—The nucleus yields from 40 to 50 per cent. of a *fatty fixed oil*, which, being official, is described below. This oil consists chiefly of the glycerides of several fatty acids ; and of *tiglinic acid*, and *crotonol.* No satisfactory evidence has yet been given of the real purgative principle of croton oil.

Oleum Crotonis.

Croton Oil.

Synonym.—Oleum Tiglii.

Production, Varieties, and Commerce.—Two varieties of croton oil are known in commerce—one which is imported from India, where it is obtained by submitting the nucleus to pressure after its removal from the coats of the slightly-roasted seeds ; and the other expressed in Britain from seeds imported from India and Cochin, after the removal of their seed-coats. The *latter oil is alone official,* and has the following characters :—

General Characters.—Brownish-yellow to dark reddish-brown, fluorescent, with a viscid consistence which is in-creased by age; a faint, peculiar, somewhat rancid, disagreeable odour, and an oily acrid taste. It is entirely soluble in alcohol.

The East Indian oil is readily distinguished by its pale-yellow colour, and by being insoluble in alcohol.

Principal Constituents.—(*See* Croton Seeds, page 302.)

Medicinal Properties.—Croton oil, when administered internally in doses of from one-third to one minim, acts as a powerful hydragogue cathartic. Externally applied it is rubefacient and counter-irritant.

Official Preparation.—Linimentum Crotonis.

5. EUPHORBIA RESINIFERA, *Berg.*

Euphorbium.

(Bentley and Trimen's ' Medicinal Plants,' vol. iv. plate 240.)

Habitat.—It is only found in the interior of Morocco, on the slopes of the Great Atlas range, in the Southern Province of Suse.

Product Used and Name.—EUPHORBIUM :—the resinous juice or gum-resin, obtained by incision in the green fleshy branches of the plant, hardened by exposure to the sun.

(*Not Official.*)

Euphorbium.

Euphorbium.   Gum Euphorbium.

Commerce.—It is collected in districts near the city of Morocco, and shipped from Mogador.

General Characters.—Euphorbium is in conical, somewhat globular, or irregular pieces or masses, which are sometimes as much as an inch across, but mostly much smaller ; these have a dull-yellow or yellowish-brown colour, waxy and translucent appearance, and have commonly mixed with them portions of the angular spiny stem of the plant from whence obtained.   The pieces are hollow, or enclose fragments of the spines or flowers ; they are brittle, have a slight aromatic odour, which is increased by heat, act as a violent sternutatory when powdered, and their taste is very acrid and burning.

Principal Constituents.— Amorphous resin, euphorbon, and mucilage or gum. The former occurs in the proportion of nearly 40 per cent., and it is on it that the acridity of euphorbium entirely depends. Euphorbon is a crystallisable resin ; it is tasteless, and is found to the extent of about 22 per cent. There is about 18 per cent. of mucilage.

Medicinal Properties.—Drastic purgative and emetic when administered internally, but its action is so violent that it is now never employed. At present it is only used externally as a rubefacient and vesicant ; and even in these respects, in Britain, its use is almost or quite obsolete, except in veterinary practice.

ORDER 10.—PIPERACEÆ.

1. PIPER ANGUSTIFOLIUM, *Ruiz & Pavon.*
Matico.

Synonym.—ARTANTHE ELONGATA, *Miquel.*

(Bentley and Trimen's ' Medicinal Plants,' vol. iv. plate 242.)

Habitat.—It is found in a wild or cultivated state over a great range in tropical America, growing in Peru, Bolivia, Columbia, Brazil, Venezuela, and elsewhere.

Official Part and Name.—MATICÆ FOLIA : —the dried leaves.

Maticæ Folia.
Matico Leaves.

Commerce.—Matico leaves appear to reach this country chiefly by way of Panama, but the plant from which they are obtained has, as already noticed, a wide range over tropical America.

General Characters.—Matico leaves, as commonly seen in commerce, are more or less broken, folded, and compressed into a brittle mass, and have mixed with them a variable proportion of jointed stems, flowers, and fruits, the whole being greenish-yellow in colour. The leaves, when entire, are very shortly petiolate, from about four to six,

X

Fig. 42 —Portion of the
stem and under sur-
face of a leaf of the
Matico plant (*Piper
angustifolium*).

or even eight inches long;
oblong-lanceolate (*fig.* 42),
tapering towards the apex,
cordate and unequal at the
base, entire or minutely
crenulate at the margins,
greenish-yellow, reticulated
with sunken veins and tessellated above, the veins
prominent beneath (*fig.* 42),
and the depressions formed
by them densely covered with
hairs. They have an aromatic
bitterish taste; and a pleasant feebly-aromatic odour.

Substitutions.—The leaves
of other plants have been imported into this country as the
official matico, but the only
spurious matico that has been
specially described is that
first noticed by the author in
1863, which is derived from
Piper aduncum, Linn.; since
which time it has been somewhat frequently found in the
London market. It may be
readily distinguished from
the official kind by being in
a less compressed state, by
the upper surface of the leaves
not being so tessellated, and
by the almost entire absence
of prominent veins and
intermediate hairs on the
under surface.

Principal Constituents.—The chief constituents are *volatile oil*, in the proportion of about 1½ per cent.; *pungent resin*; a crystallisable acid called *artanthic acid*; and a little *tannic acid.* Its medicinal properties appear to be essentially due to the volatile oil and resin, and also, to some extent, to the tannic acid.

Medicinal Properties.—Mild aromatic tonic, and stimulant, acting more like cubebs and pepper on the genitourinary organs and rectum. When locally applied it acts as a styptic; it is also frequently regarded as an internal styptic, or hæmostatic, and in Peru it is used as an aphrodisiac.

Official Preparation.—Infusum Maticæ.

2. PIPER CUBEBA, *Linn. fil.*
Cubebs.

Synonym.—CUBEBA OFFICINALIS, *Miquel.*

(Bentley and Trimen's ' Medicinal Plants,' vol. iv. plate 243.)

Habitat.—It is indigenous in Java, Sumatra, and Borneo.
Official Parts or Products and Names.—1. CUBEBA :—the dried unripe full-grown fruit. 2. OLEUM CUBEBÆ :—the oil distilled in Britain from cubebs.

1. Cubeba.
Cubebs.

Cultivation and Commerce.—The Cubeb plant is cultivated in Java and Sumatra; but Singapore is the great emporium from which cubebs are shipped to Europe and elsewhere.

General Characters and Test.—Cubebs resemble black pepper in size and form, but are at once distinguished by each fruit being furnished with a stalk or pedicel, hence their common name of *tailed pepper.* Cubebs are globular, from about one-sixth to one-fifth of an inch in diameter,

X 2

blackish or greyish-brown, much wrinkled on the surface, and tapering below into a rounded stalk, which is continuous with, and permanently attached to, the pericarp. This stalk is commonly a little longer than the fruit itself, or, rarely, it is nearly twice as long. Beneath the shrivelled skin is a hard brown smooth shell (*endocarp*) in which there is a solitary seed in the mature fruit, but in commercial cubebs this seed is mostly so little developed that the endocarp is nearly empty. The taste is warm, aromatic, and somewhat bitter; and the odour strong, peculiar, and aromatic. A decoction when cold is coloured bright indigo-blue by solution of iodine.

Substitutions and Adulterations.—In consequence of the enormous demand of late years, cubebs are now frequently imported of inferior quality, and often mixed more or less with inert stalks or peduncles, or other fruits are substituted for them. Two of these spurious fruits have been specially described lately, and must be briefly referred to. The more important is said to be derived from *Piper crassipes*, Korthals, a Sumatran species. This spurious fruit is distinguished by its somewhat larger size, lighter colour, stouter and flattened stalk, somewhat mace-like odour, more bitter taste, and by its decoction when cold being coloured dull purplish instead of indigo-blue with solution of iodine. The other spurious fruit has been referred to *Daphnidium Cubeba*, Nees, a Lauraceous plant. It may be known by its less wrinkled skin, pleasanter taste, by its contained seed readily splitting into two oily cotyledons, and by its decoction when cold presenting no change of colour on the addition of solution of iodine.

Principal Constituents.— *Volatile oil* (which is described below); a neutral crystalline substance, termed *cubebin* ; an acid amorphous resin (*cubebic acid*); and a *neutral amorphous resin.* The medicinal properties have been generally ascribed to the volatile oil, but more recent observations seem to prove that the peculiar effects of cubebs are due to the neutral resin and cubebic acid. Cubebin appears to be inert.

2. Oleum Cubebæ.

Oil of Cubebs.

Production.—The yield of oil varies much, namely from 5 to 15 per cent., according to the qualities of the cubebs, and the temperature employed in their distillation, etc. It is directed to be distilled in Britain.

General Characters.—Colourless or pale greenish-yellow, with the odour and taste of cubebs in a concentrated degree. It is levogyrate, and has a specific gravity of about 0·929.

Medicinal Properties of Cubebs and Oil of Cubebs.— Cubebs have stimulant and diuretic properties; their stimulant effects being more especially manifested on the genitourinary mucous membranes. Oil of cubebs has also been supposed to possess all the essential properties of cubebs, but recent investigations, as already noticed, indicate that the special efficacy of cubebs depends entirely upon the acid resin (*cubebic acid*) and the neutral resin—the volatile oil being simply stimulant and carminative.

Official Preparations of Cubebs.

Oleo-resina Cubebæ.   *Dose.*—5 to 30 minims.
Oleum Cubebæ.   *Dose.*—5 to 20 minims.
Tinctura Cubebæ.   *Dose.*—½ to 2 fluid drachms.

---

3. PIPER NIGRUM, *Linn.*

Black Pepper.

(Bentley and Trimen's ' Medicinal Plants,' vol. iv. plate 245.)

Habitat.—It is a native of Southern India, especially the Malabar Coast, whence it has been introduced into Java, Borneo, Sumatra, the Malay Peninsula, Siam, the Philippines, and the West Indies. Its cultivation is carried on more or less in all these countries, but especially in Southern India.

Official Part and Name.—PIPER NIGRUM :—the dried unripe fruit.

Piper Nigrum.

Black Pepper.

Preparation, Commerce, and Varieties.—The fruits are gathered when those at the base of the spike begin to change colour from green to red ; they are then separated from the stalk or rachis by hand rubbing, picked clean, and subsequently dried by exposure to the sun, or more commonly by the heat of a gentle fire.   Singapore is the great emporium for black pepper, but some comes from British India and other parts.   There are several varieties known, as Malabar, Cochin, Penang, Singapore, and Siam ; Malabar being the most esteemed.

General Characters.—Black pepper is roundish, and usually about one-fifth of an inch in diameter. The pericarp is thin, wrinkled, and of a blackish-brown colour; it contains a solitary hard smooth roundish seed, of a yellowish-brown or grey colour externally, and mealy within.   It has an aromatic odour, and a pungent somewhat bitterish taste.   The heaviest fruits are the most valued, and are called *shot pepper.*

Principal Constituents.—The principal constituents of black pepper are *resin* ; *volatile oil*; and *piperine* ; and the same substances are also present in white pepper (*see below*), but in somewhat different proportions. The odour and aromatic taste are due to the volatile oil; and the pungent taste to the resin. Piperine is a feebly alkaline crystalline substance, and to its presence the febrifugal properties of pepper are due.

Medicinal Properties.—Aromatic, carminative, stimulant, and somewhat febrifugal when administered internally; its stimulant action being more especially evident on the mucous membranes of the rectum and genito-urinary organs.   Externally applied it is rubefacient.

Official Preparations.

Pulvis Opii Compositus, which   of Confectio Opii.
is used for the preparation   Confectio Piperis.

White Pepper.

(Not Official.)

White Pepper is prepared from the ripe fruits of the black pepper plant, as follows :—After the spikes of ripe fruit have been gathered two or three days, they are washed and bruised in a basket with the hand till all the stalks and pulp of the pericarps are removed, after which they are dried. Hence while black pepper is the dried unripe entire fruit, white pepper is obtained from the ripe fruit, and is little more than the seed. White pepper is somewhat larger than black pepper, smooth, roundish, greyish or yellowish-white; hard and horny externally, and with a similar taste and odour to black pepper, but less pungent. Its constituents are the same as those of Piper Nigrum.

---

4. PIPER LONGUM, *Linn.*

Long Pepper.

Synonym.—CHAVICA ROXBURGHII, *Miquel.*

(Bentley and Trimen's ' Medicinal Plants,' vol. iv. plate 244.)

Habitat.—It is found in many parts of Southern and Eastern India, especially the Malabar and Coromandel coasts, where it is also largely cultivated. It grows also in Ceylon, Timor, and the Philippine Islands.

Part Used and Name.—PIPER LONGUM :—the dried spike of unripe fruits of the above species ; and of Piper officinarum, *DC.*

(Not Official.)

Piper Longum.

Long Pepper.

Preparation and Commerce.—Long pepper is gathered when the spikes of fruits are full grown but yet unripe ; and then dried by exposure to the sun. It is chiefly exported from Singapore and Penang, where it has been principally derived from Java, but also, to a much smaller extent, from Rhio—that from Java being obtained in a great measure from *Piper officinarum*, DC. It is also largely exported from Calcutta.

Characters.—Long pepper is cylindrical in form, and from about one inch to an inch and a half in length, and about a quarter of an inch in thickness. It consists of a number of sessile fruits closely arranged on a common axis (*rachis*), and has a brownish-white colour from having been rubbed in lime or some other earthy powder. It tapers somewhat above, and is marked with spirally arranged superficial furrows. The odour is agreeable and somewhat aromatic ; and its taste very pungent and aromatic.

Principal Constituents.—Same as those of black pepper.

Medicinal Properties.—Its medicinal properties are similar to those of black pepper; but the latter has now almost entirely replaced it in medicine, except in veterinary practice.

SERIES 2.—*INFERÆ* OR *EPIGYNÆ.*

ORDER I.—ARISTOLOCHIACEÆ.

1. ARISTOLOCHIA SERPENTARIA, *Linn.*
Virginian Snakeroot.

(Bentley and Trimen's ' Medicinal Plants,' vol. iv. plate 246.)

2. ARISTOLOCHIA RETICULATA, *Nutt.*
Texan Snakeroot.   Red River Snakeroot.

Habitat.—1. ARISTOLOCHIA SERPENTARIA.—A native of the United States of America, growing in all except the most Northern States. 2. ARISTOLOCHIA RETICULATA.—It is indigenous to Louisiana and Texas.

Official Part and Name.—SERPENTARIÆ RHIZOMA :— the dried rhizome and rootlets of the above two species.

Serpentariæ Rhizoma.

Serpentary Rhizome.

Synonym.—Serpentariæ Radix.

Collection and Commerce.—Virginian Snakeroot or Middle States Snakeroot is collected in the States on the east of the Mississippi, or in the middle States; and the Red River or Texan Snakeroot is obtained from the States south-west of the Rocky Mountains. Serpentary rhizome is imported from the United States in bales, casks, or bags.

General Characters.—Rhizome twisted, about one inch long and one-eighth of an inch in diameter ; marked above by the short remains of former aerial stems, and giving off from its under surface numerous slender branched rootlets of from two to four inches long, which form together a more or less interlacing tuft. It has a dull yellowish-brown colour externally, and is whitish within; a peculiar, aromatic, some-what camphoraceous odour; and a bitterish, aromatic, camphoraceous taste.

The rhizomes and rootlets of the two species of *Aristolochia* have a marked resemblance, but the rhizome of *Aristolochia reticulata* is a little thicker ; and the rootlets are longer, coarser, and less matted together than the corresponding parts of *A. Serpentaria.* It is also said to be somewhat less aromatic. When any of the leaves of the plant are present in the drug its botanical source is at once evident, as those of *A. reticulata* are evidently reticulated on their under surface, and somewhat coriaceous in texture.

Adulterations and Substitutions.—The roots and rhizomes of other plants are not unfrequently substituted for, or mixed with, serpentary rhizome, either fraudulently or by accident ; such as the roots of *Spigelia marilandica*, Linn., and those of *Panax quinquefolium*, Linn.; and the rhizomes and rootlets of *Cypripedium pubescens*, Linn., and those of *Hydrastis canadensis*, Linn. These, however, are readily distinguishable by the characters given above, if only ordinary care be taken.

Principal Constituents.—*Volatile oil* in the proportion of about ½ per cent., to which the odour of serpentary rhizome is due; and a *bitter principle* (*aristolochin*), which requires further examination. A little *tannic acid, resin,* and some *starch,* etc., are also present.

Medicinal Properties.—Stimulant, tonic, diaphoretic, and diuretic ; but it has no remedial value, as its common name implies, for the cure of the bites of venomous reptiles, or in the bites of mad dogs as also sometimes stated.

Official Preparations.

Infusum Serpentariæ. | Tinctura Cinchonæ Composita.
Tinctura Serpentariæ.

ORDER 2.—CORYLACEÆ OR CUPULIFERÆ.

1. QUERCUS ROBUR, *Linn.*

Common Oak.

Synonym.—QUERCUS PEDUNCULATA, *Ehr.*

(Bentley and Trimen's 'Medicinal Plants,' vol. iv. plate 248.)

Habitat.—A very common tree in Britain and other parts of Europe, and also in Western Asia, extending far north, but not found in the Arctic regions. It also grows in Syria and North Africa.

Official Part and Name.—QUERCUS CORTEX :—the dried bark of the smaller branches and young stems.

Quercûs Cortex.

Oak Bark.

Collection and Preparation.—Oak bark is directed in the British Pharmacopœia to be collected in the spring, from trees growing in Britain. This direction should be carefully followed, because at this season the bark contains the larger proportion of tannic acid ; but as a general rule, the trees are barked from the beginning of May to about the middle of July. After the bark is separated, it is dried in the air by setting it on what are called lofts or ranges, and then stacked.

General Characters.—Oak bark occurs in quills of varying lengths and sizes, the bark itself being commonly about one-tenth of an inch or less thick. It is covered with a smooth, shining, silvery or ash-grey corky layer, variegated with brown, and dotted over with small scars. Internally it is brownish-red or cinnamon-brown and longitudinally striated. Its fracture is tough and fibrous ; its taste very astringent ; and its odour very feeble, although when moistened it resembles tan.

Principal Constituents.—The most important constituent is a form of tannic acid (*querci-tannic acid*), which occurs in the proportion of from 7 to 10 per cent. when the bark is collected in the spring. It also contains a small quantity of *gallic acid.*

Medicinal Properties.—Astringent. It is chiefly used for external application.

     Official Preparation.—Decoctum Quercûs.

---

2. QUERCUS LUSITANICA, *Webb*, var. INFECTORIA, *DC.*

Aleppo Galls. Turkey Galls.

Synonym.—QUERCUS INFECTORIA, *Oliv.*

(Bentley and Trimen's ' Medicinal Plants,' vol. iv. plate 249.)

Habitat.—It occurs abundantly in Syria and Asia Minor. It is also found in Greece, Southern Turkey, and Cyprus.

Official Part and Name.—GALLA :—excrescences on the

above plant, caused by the puncture and deposit of an egg
or eggs of Cynips Gallæ tinctoriæ, *Oliv.*

Galla.

Galls.

Synonym.—Nutgalls.

Production.—The official galls are morbid excrescences
or tumours of hypertrophied vegetable tissue formed on the
above plant—the result of the puncture by the hórny ovipo-
sitor of the female hymenopterous insect (*Cynips Gallæ
tinctoriæ*), and the deposit of one or more eggs. When the
gall is fully developed the egg is hatched, and the larva or
grub begins to feed on the juices of the central tissue, and
afterwards passes into the pupa or chrysalis stage, and ultimately into a small four-winged fly, which bores a cylindrical
passage through the gall with its mandibles, and escapes.
The best galls are those which are gathered while the larva
is undergoing its transformations—that is, before the insect
has escaped.

Varieties and Commerce.—Galls or nutgalls are commonly
known in commerce as Aleppo, Turkey, or Levant galls ;
and are exported from various Turkish ports, and also from
Persia. Two varieties are especially distinguished, namely,
blue or *green galls*, and *white galls*, the former of which are
the most esteemed, being gathered before the escape of the
insect, and *are alone official.*

General Characters.—Hard, heavy, sub-globular, from
about one-half to three-quarters of an inch or more in
diameter, more or less tuberculated on the surface, the
tubercles and intervening spaces being smooth; dark bluish-
green or dark olive-green externally, yellowish or brownish-
white internally, with a small central cavity in which may
be found the remains of the larva or more or less developed insect, according to the period at which the galls
have been gathered. They have a somewhat granular frac-

ture; no marked odour; but an intensely astringent taste, followed by some degree of sweetness.

White Galls.—These galls are larger, lighter-coloured (being yellowish or whitish), less compact, less heavy, less astringent, perforated with a circular hole, and having a larger central cavity with no remains of the insect. They are *not official.*

Principal Constituents.—The best official galls yield from 60 to 70 per cent. of *tannic acid (gallo-tannic acid)* ; and about 3 per cent. of *gallic acid* ; and also some *sugar, starch,* &c.

Medicinal Properties.—Galls are the most powerful of all vegetable astringents, and hence are applicable in all cases where astringent medicines are required, both for external and internal use. For internal use, however, the official tannic and gallic acids, which are obtained from them, are much more largely employed.

Official Preparations.

Acidum Gallicum, which is also used in the prepara-tion of Glycerinum Acidi Gallici.

Acidum Tannicum, which is also used in the preparation of Glycerinum Acidi Tan-nici ; Suppositoria Acidi Tannici; Suppositoria Acidi Tannici cum Sapone ; and Trochisci Acidi Tannici.

Tinctura Gallæ.

Unguentum Gallæ, which is also employed in the pre-paration of Unguentum Gallæ cum Opio.

ORDER 3.—SANTALACEÆ.

SANTALUM ALBUM, *Linn.*

Sandal Wood.   Yellow Sanders Wood.

(Bentley and Trimen's ' Medicinal Plants,' vol. iv. plate 252.)

Habitat.—It is indigenous to Mysore and the neigh-bouring parts of Southern India. It is also found in the islands of the Eastern Archipelago, especially Sumba or Sandal-wood Island, and Timor.

Official Product and Name.—OLEUM SANTALI :—the oil distilled from the wood.

Source and General Characters of Sandal Wood.—The best sandal wood is obtained from the Southern part of India, and is known as Malabar Sandal Wood ; but some is also derived from the Islands of the Malay Archipelago. According to Umney, all the sandal wood imported into this country now comes from Bombay. Sandal wood is also obtained from other species of *Santalum*, but all so derived is of inferior quality. The Malabar kind is in billets or logs from which the bark and sapwood have been removed, and varying in length from two to four feet, and in diameter from three to eight inches. It is hard, heavy, yellowish-brown, with darker-coloured zones in a transverse section, and with very narrow medullary rays. When ground, rasped, or rubbed, it has a very agreeable aromatic odour, somewhat rose-like; and its taste is strongly aromatic and bitterish.

Oleum Santali.

Oil of Sandal Wood.

Synonym.—Oleum Santali Flavi.

Production.—This volatile oil is procured from the heartwood after it has been cut into small chips, by distillation with water. The amount varies much, probably according to the kind of sandal wood from whence obtained ; thus, in 'Pharmacographia,' the yield is said to be about 1 per cent., but Bidie gives it at 2·5 per cent. ; Umney has, however, obtained as much as 4·5 per cent., but sometimes less than one-third of this quantity will result even after careful distillation. Bidie also states that the largest quantity of oil, and that of the best quality, is derived from the roots.

General Characters.—Oil of sandal wood has a thick

consistence, pale-yellow colour, a strongly aromatic odour, a pungent and spicy flavour, and is neutral or slightly acid in reaction. It specific gravity is usually about 0·963, but this, as well as the quality of the oil, varies much. W. H. Ince found the specific gravity of different samples vary from 0·9713 to 0·9797. It is readily soluble in alcohol.

Medicinal Properties.—Stimulant and sudorific, its stimulant action being more especially evident on the mucous membranes of the genito-urinary organs, hence it has acquired some reputation as a substitute for copaiba and cubebs in gonorrhœa, etc., and is said sometimes to have succeeded when both these remedies had previously failed. Its pleasant taste and odour also recommend it.

Dose.—10 to 30 minims.

Class II.　MONOCOTYLEDONES.

Sub-class I.—PETALOIDEÆ.

Series 1.—*INFERÆ* or *EPIGYNÆ*.

Order 1.—MARANTACEÆ or CANNACEÆ.

MARANTA ARUNDINACEA, *Linn.*

Arrowroot.

(Bentley and Trimen's ' Medicinal Plants,' vol. iv. plate 265.)

Habitat.—It is a native of the tropical parts of America from Mexico to Brazil, and of the West Indian Islands. In the old world it grows in Bengal, Java, the Philippines, etc., either wild or cultivated ; and in Mauritius, at Natal, and on the west coast of Africa. It is also largely cultivated in parts of Brazil, in Georgia, and in the Bermudas and other West Indian Islands, Queensland, etc.

Product Used and Name.—MARANTA :—the starch obtained from the rhizome.

(*Not Official.*)

Maranta.

Arrowroot.   Maranta Starch.

Synonym.—Marantæ Amylum.

Extraction and Commerce.—In Jamaica the starch is extracted from the scraped and washed rhizomes when about a year old, by reducing them to a pulp, which is afterwards well stirred and washed in cold water (usually on sieves), and the water which passes through holding the starch in suspension is collected in suitable vessels, and allowed to stand until the starch is deposited at the bottom, after which the water is poured off, and the starch is then dried in the sun.   In all stages of its preparation great care is taken, so that maranta starch is commonly a very pure product.   It is imported from several of the West Indian Islands, as Bermuda, St. Vincent, Jamaica, etc.   Bermuda is regarded as the best variety, but it is now produced to a very limited extent.   Maranta starch is also produced at Sierra Leone, Natal, Brazil, East Indies, etc.

General Characters.—Arrowroot or maranta starch is white, opaque, odourless, and tasteless; and is either found in

FIG. 43.   Maranta Starch or Arrowroot (magnified 250 diameters).

the form of small aggregated masses, which, when rubbed, emit a faint crackling sound, or in fine powder. Examined by the microscope, the granules which have a glistening appearance, have been thus described by Pereira : 'Convex, more or less elliptical, and moderately uniform in size (*fig.* 43). Their shape is more or less irregular, but often oblong, or usually somewhat ovate-oblong, frequently obscurely triangular, or oyster-shaped, or mussel-shaped. The rings are very evident, though fine. The nucleus or hilum is usually most distinct, and generally placed towards one end of the granule; it is normally circular, but frequently cracked in a linear or stellate manner.'

Composition.—Similar to that of other pure starches.

Adulterations and Substitutions.—Arrowroot, from its high price, is frequently adulterated with cheaper starches, or these are substituted for it. The microscope is the only means by which these frauds can be satisfactorily detected.

Medicinal Properties.—Demulcent and emollient; and hence valuable in bowel complaints and diseases of the urinary organs, etc.

ORDER 2.—ZINGIBERACEÆ OR SCITAMINACEÆ.

1. ELETTARIA CARDAMOMUM, *Maton.*

Malabar Cardamom.

(Bentley and Trimen's 'Medicinal Plants,' vol. iv. plate 267.)

Habitat.—This plant is a native of Southern India, growing abundantly in the hills of North Canara, Coorg, and Wynaad. It is also cultivated in the same districts.

Official Part and Name.—CARDAMOMI SEMINA :—the dried ripe seeds.

Cardamomi Semina.

Cardamoms.

Production and Preparation.—Cardamoms are produced in Southern India, and are chiefly derived from cultivated plants, but to some extent also from those growing wild. The fruits, after collection, are either dried on their flower-stalks, or after being removed from them, and either wholly

Y

by sun-heat, or first by sun-heat and subsequently by a gentle fire-heat. The seeds are best kept in their pericarps, in which condition they are imported; but when required for use, as directed in the British Pharmacopœia, they should be separated and the pericarps rejected.

Commerce and Varieties.—Cardamoms are commonly brought to Europe by way of Bombay, and are known in commerce by the districts from whence they are obtained, as *Malabar*, *Madras*, and *Aleppy*, the former being the most esteemed.

General Characters.—The dried fruit, as seen in commerce, is ovoid or oblong in form, three-sided, three-celled, shortly beaked at the apex and rounded at the base, where the remains of the stalk may frequently be found. The fruits vary in length from about two-fifths to nearly an inch, and in breadth from about one-fifth to two-fifths of an inch. The pericarp is obtusely triangular, brownish-yellow, and longitudinally striated externally; thin, somewhat coriaceous, and encloses from about fifteen to twenty seeds. The seeds, *which are alone official*, constitute about seventy-five parts by weight of the whole; they are about one-sixth of an inch long, irregularly angular or somewhat wedge-shaped, transversely wrinkled, dark reddish- or yellowish-brown externally, with a small depressed hilum; and whitish within. The pericarp has no marked odour or taste; but the seeds have an agreeable aromatic odour, and a warm aromatic taste.

In trade, cardamoms are distinguished as *shorts* and *short-longs*. The *shorts* are plump, heavy, ovoid or somewhat globular in form, and from about two-fifths to three-fifths of an inch in length; the *short-longs* are more oblong and tapering at each end, paler in colour, and of greater length, varying in this respect from about seven-tenths to nearly one inch. The shorts are the most esteemed.

Principal Constituents.—Cardamoms owe their properties to a *volatile oil*, which is contained in the proportion of from about 4 to 5 per cent. This oil has the odour and taste of the seeds, is strongly dextrogyre, colourless when fresh, but becoming yellowish by keeping, and has a specific gravity of about 0·93. They also contain about 10 per cent. of a *fixed oil.*

Medicinal Properties.—Carminative, aromatic, and stimulant. They are used largely, on account of their agreeable aromatic properties, as an adjunct to other medicines to improve their flavour, and as correctives.

Official Preparations.

Extractum Colocynthidis Compositum.

Pulvis Cinnamomi Compositus, which is an ingredient in Pilula Aloes et Ferri; and Pilula Cambogiæ Composita.

Pulvis Cretæ Aromaticus.

Tinctura Cardamomi Composita, which is also contained in Decoctum Aloes Compositum; Mistura Ferri Aromatica; Mistura Sennæ Composita; and Tinctura Chloroformi Composita.

Tinctura Gentianæ Composita.

Tinctura Rhei.

Vinum Aloes.

OTHER KINDS OF CARDAMOMS.

(*Not Official.*)

Besides the above official cardamoms, derived from the plant commonly distinguished as the Malabar Cardamom, several other closely allied plants yield seeds which have been, or are now, more or less employed in pharmacy and for other purposes. They are known in commerce as Ceylon Cardamoms, Round Cardamoms, Bengal Cardamoms, etc., but for a description of these we must refer to larger works on the Vegetable Materia Medica. The seeds, however, known commonly as Grains of Paradise (Grana Paradisi), and which are derived from an allied plant, require from us a short notice.

Grains of Paradise.
(*Not Official.*)

Botanical Source and Commerce.—*Grains of Paradise, Guinea Grains,* or *Melegueta Pepper,* are obtained from the Gold Coast, in Western Tropical Africa, and are derived from Amomum Melegueta, *Roscoe*; Bentley and Trimen's 'Medicinal Plants,' vol. iv. plate 268.

General Characters.—These seeds are from about one-eighth to one-tenth of an inch long, roundish-ovoid or somewhat wedge-shaped in form, and bluntly angular. They are wrinkled, pitted, and somewhat warty externally, of a shining golden-brown or reddish-brown colour, and with a paler coloured and somewhat beak-shaped hilum; they have a hard texture, and when broken are whitish internally. Their odour is slightly aromatic; and their taste very pungent and pepper-like.

Principal Constituents.— *Volatile oil* and *acrid resin,* to which their properties are due. According to Thresh a similar pungent principle exists to that he found in capsicum fruit (*capsaicin*), and which he has named *paradol.*

Medicinal Properties.—Stimulant, like the peppers. They are essentially used in this country in the preparation of cattle medicines, and to give pungency to spirits.

---

2. CURCUMA LONGA, *Linn.*
Turmeric.

(Bentley and Trimen's 'Medicinal Plants,' vol. iv. plate 269.)

Habitat.—Its native country cannot now be determined, but probably some part of the Indian peninsula.

Official Part and Name.—TURMERIC:—the dried rhizome.

Curcuma.

Turmeric.

Cultivation, Commerce, and Varieties.—Turmeric is obtained from plants cultivated for the purpose in various parts of Southern Asia, more especially about Calcutta and throughout Bengal. It is distinguished in commerce from the countries or districts whence obtained, as *China, Bengal, Madras, Java,* etc.; the Chinese is the most esteemed, but is rarely met with; the Java is of low commercial value.

FIG. 44.—Madras Long Turmeric.   FIG. 45.—Madras Round Turmeric.

General Characters.—The turmeric of commerce consists of two forms of rhizomes or tubers—the *round* (*fig.* 45), and the *long* (*fig.* 44), the former being the central rhizomes, and the latter the lateral ones. The *round* rhizomes (*fig.* 45), when entire, are from one and a half to two inches long, roundish or somewhat ovoid, pointed at one end, and marked with annular ridges; they are often found cut in halves.

The *long* or *lateral* rhizomes (*fig.* 44) are from two to three inches long, sub-cylindrical, somewhat curved, pointed at both ends, about the thickness of the little finger, and marked on the surface with annular ridges, and frequently having one or more little projections or shoots on their sides. Both kinds are dark yellowish-brown externally, very hard and firm, and internally having a dull waxy-resinous appearance and an orange-yellow or reddish-brown colour; the powder is deep orange-yellow. Turmeric has an aromatic peculiar taste and odour; and when chewed it tinges the saliva yellow.

Principal Constituents.—*Volatile oil*, to which its odour is due, a little *pungent resin, starch*, and a yellow colouring matter (*curcumin*). It is from the presence of curcumin that turmeric is introduced into the Appendix of the British Pharmacopœia as an article for employment in chemical testing, as the alkalies change its colour to brownish-red, and boric or boracic acid produces with it an orange-red tint.

Medicinal Properties.—Stimulant and carminative, like ginger, but it is not now used medicinally, but only as a condimentary substance in curry powder, etc. As already stated, it is introduced into the Appendix of the British Pharmacopœia for use as a test agent.

Official Preparations.

Turmeric Paper. | Tincture of Turmeric.

---

3. ZINGIBER OFFICINALE, *Roscoe.*

Ginger.

(Bentley and Trimen's ' Medicinal Plants,' vol. iv. plate 270.)

Habitat.—It is not known in a truly wild state, but is undoubtedly a native of Tropical Asia. From Asia it was carried to the West Indies, where it is now abundant, and is indeed cultivated in all the warmer regions of both hemispheres.

Official Part and Name.—ZINGIBER :—the scraped and dried rhizomes.

Zingiber.

Ginger.

Preparation and Commerce.—Ginger is seen in commerce in two forms depending upon its mode of preparation, which are distinguished as *uncoated* or *scraped ginger* (*white ginger*), and *coated* or *unscraped ginger* (*black ginger*). The latter is prepared by simply drying the washed rhizomes when about a year old by the heat of the sun ; and the former are dried after the removal of their epidermal integument or skin. The latter is regarded as the best, and is *alone official.* Several varieties of ginger are distinguished in commerce, as Jamaica, Cochin, Bengal, and African ; the African being coated, the other three uncoated. The Jamaica is regarded as the best variety, and next to it the Cochin.

General Characters.—The official rhizomes are in flattish irregularly-branched pieces, which are commonly termed races or hands; these vary in length, but are usually from about three to four inches, and each branch is marked at its summit by a depressed scar, which indicates the point of attachment of a former flowering stem or scape. Externally the rhizome has a pale buff colour, and is somewhat fibrous and striated ; it breaks readily with a mealy short fracture, but with some projecting fibres. *Coated* or *unscraped* ginger is at once distinguished from the above by being covered with a wrinkled brown or brownish-black integument ; and internally it is also usually harder and darker coloured. Ginger has an agreeable aromatic odour, and a strong pungent taste.

Principal Constituents.—*Starch, resin,* and *volatile oil,* the odour being due to the volatile oil, and its pungent taste to the resin. According to Thresh, a similar pungent principle exists in ginger to that he found in capsicum fruit (*capsaicin*) ; which he has termed *gingerol.*

Medicinal Properties.—Stimulant, aromatic, and carminative, when taken internally ; sialagogue when chewed ; and externally applied it is rubefacient.

Official Preparations.

Confectio Scammonii.

Infusum Sennæ, which is also used in the preparation of Mistura Sennæ Composita.

Pilula Scillæ Composita.

Pulvis Cinnamomi Compositus, which is an ingredient in Pilula Aloes et Ferri ; and Pilula Cambogiæ Composita.

Pulvis Jalapæ Compositus.

Pulvis Opii Compositus, which is used in the preparation of Confectio Opii.

Pulvis Rhei Compositus.

Pulvis Scammonii Compositus.

Syrupus Zingiberis.

Tinctura Zingiberis.

Tinctura Zingiberis Fortior, which is also contained in Acidum Sulphuricum Aromaticum ; Pilula Scammonii Composita ; and Syrupus Zingiberis.

Vinum Aloes.

ORDER 3.—IRIDACEÆ.

1. IRIS FLORENTINA, *Linn.*

White Flag.

(Bentley and Trimen's ' Medicinal Plants,' vol. iv. plate 273.)

Habitat.— It is truly wild in the coast region of Macedonia, and the south-west shores of the Black Sea ; it is also found in several other parts of Southern and Eastern Europe. In the neighbourhood of Florence and Lucca it also occurs, but, according to Hanbury, only as a naturalised plant.

Part Used and Name.—IRIS :—the peeled, trimmed, and dried rhizome.

(*Not Official.*)

Iris.

Orris Rhizome.

Synonym.—Iridis Rhizoma.

Source, Preparation, and Commerce.—Orris rhizome, or orris root as it is more commonly, but incorrectly, termed, is obtained indiscriminately, in Tuscany, from three species of *Iris*, namely, *I. florentina*, Linn., *I. germanica*, Linn., and *I. pallida*, Lam., all of which are known under the common name of *Giaggiolo*. The rhizomes are dug up in the latter part of the summer, and afterwards peeled, trimmed, and dried in the sun, and then separated according to their qualities into *selected* and *sorts*.

General Characters. — In flattened somewhat curved branched pieces, of from two to four inches in length; these are marked on the upper surface with transverse lines, and on the lower with numerous small round brownish scars, which indicate the points where the rootlets have been cut off. They have a dull whitish colour; a compact texture; a short, somewhat mealy fracture; a bitterish, slightly acrid, and faintly aromatic taste; and an agreeable violet-like odour.

Principal Constituents.—The principal constituent is a kind of stearoptene termed *orris camphor*, which, according to the authors of ' Pharmacographia,' is simply *myristic acid* impregnated with a little *volatile oil*. *Starch* ; *acrid resin* ; and *tannic acid*, are also constituents.

Medicinal Properties.—Cathartic and emetic, and, when powdered, errhine. In Britain, at the present day, it is scarcely ever used except as an ingredient in tooth-powders and in perfumery.

2. CROCUS SATIVUS, *Linn.*
Saffron.

(Bentley and Trimen's ' Medicinal Plants,' vol. iv. plate 274.)

Habitat.—The Saffron plant is probably a native of Greece and Asia Minor, and perhaps also of Southern Italy and Persia, but it has been so long under cultivation that it is now difficult to say where it is truly wild.

Official Part and Name.—CROCUS :—the dried stigmas and top of the style.

Crocus.
Saffron.

Collection, Preparation, and Commerce.—At the present day saffron is chiefly collected for commercial purposes from plants cultivated in France, Italy, and Spain ; the varieties are known as French, Italian, and the two from Spain as Alicante and Valencia Saffron. The flowers are first collected and taken home, the stigmas with the top of the style (*fig.* 46, *b*) are then removed, and carefully dried, usually over a gentle fire. It is said that from 7,000 to 8,000 flowers only yield $17\frac{1}{2}$ ounces of fresh saffron, which by drying is reduced to $3\frac{1}{2}$ ounces.

General Characters.—The official saffron, sometimes known as *hay saffron,* as seen in commerce, is a loosely entangled mass of the dried stigmas and tops of the styles, which are either in entire portions or more or less separated into their constituent parts. Each entire portion of this mass of saffron is an inch or somewhat more in length ; it consists of three thread-like (*fig.* 46, *b*) deep orange-red stigmas, which are thickened and tubular (*fig.* 46, *a*) above, and jagged or notched at their extremities, and united below to the top of the narrow yellow style. Saffron is flexible, unctuous to the touch, with a peculiar penetrating aromatic odour, and a bitter somewhat aromatic taste. When chewed it tinges the saliva deep yellow; and when rubbed on the wet finger it produces an intense orange-yellow stain.

Tests.—When pressed between folds of white filtering paper it leaves no oily stain. When a small portion is placed in a glass of warm water it colours the liquid orange-yellow, but should not deposit any white or coloured powder. Ignited with free access of air, it yields about 6 per cent. of ash. There should not be more than 10 per cent. of moisture present.

Fig. 46. *b.* Upper part of the flower of *Crocus sativus*, laid open to show the style and three stigmas. *a.* One of the stigmas, *enlarged.*

Adulterations.—Saffron, from its high price, is very frequently adulterated. Thus, to give it freshness, flexibility, and increased weight, it is damped or oiled or dipped in glycerine. It is often mixed with the florets of the Marigold (*Calendula officinalis*) dyed with logwood, or of those of the Safflower (*Carthamus tinctorius*), or of *Arnica montana*, and other plants ; or with strips of the petals of different plants ; or with the dyed stamens of the saffron crocus ; or with the coloured fibres or shreds of hard boiled or smoked beef ;

and in various other ways. All these frauds are readily detected by putting into a glass of warm water a few shreds of the suspected drug, when these admixtures may be known by their different appearances from the marked structure of the genuine drug.

Of late years saffron has been frequently adulterated by being coated with coloured chalk, sulphate of lime, and other earthy matters. This fraud is at once made evident by stirring the suspected drug in a glass of water, when, if so adulterated, the water will become turbid, and ultimately deposit the white or coloured powder at the bottom of the vessel.

Principal Constituents.—Saffron yields by distillation with water about 1 per cent. of a *volatile oil*, to which its odour is due; but the principal constituent is an orange-red colouring matter termed *polychroit*, a glucoside, which when decomposed yields *sugar*; a red colouring matter called *crocin*; and a *volatile oil* having the odour of saffron, and probably identical with that obtained by distillation.

Medicinal Properties. -- Saffron has been regarded as stimulant, antispasmodic, and emmenagogue; but at present it is scarcely ever employed in medicine except as a colouring and flavouring agent.

Official Preparations.

Decoctum Aloes Compositum.
Pilula Aloes et Myrrhæ.
Pulvis Cretæ Aromaticus.
Tinctura Cinchonæ Composita.
Tinctura Croci.
Tinctura Opii Ammoniata.
Tinctura Rhei.

SERIES 2.—*SUPERÆ.*

ORDER 1.—LILIACEÆ.

1 URGINEA SCILLA, *Steinheil.*
Squill.

(Bentley and Trimen's 'Medicinal Plants,' vol. iv. plate 281.)

Habitat.—It occurs in most parts of the Mediterranean district, especially on the sea coast, as in Southern France, Portugal, Southern Spain, Italy, Dalmatia, Greece, Syria, Asia Minor, North Africa, and the Mediterranean Islands. It is also found at the Cape of Good Hope and the Canary Islands.

Official Part and Name.—SCILLA :—the bulb; divested of its dry membranous outer scales, cut into slices, and dried.

Scilla.
Squill.

Varieties, Preparation, and Commerce.—Squill bulb has the ordinary structure of a tunicated bulb, but two varieties are distinguished—the *white* and the *red.* The former is so termed because the outer and inner scales are all colourless; while in the latter the inner fleshy scales are of a pale rose tint, and the external dry membranous ones reddish-brown. For use in medicine, the bulbs are first freed from their outer membranous scales, and then cut transversely into slices ; these are then usually dried in the sun, and subsequently packed in casks for exportation, or very rarely, the bulbs are forwarded in their entire state. Our supplies are derived from Malta. As they greedily absorb moisture, they should be carefully preserved in a dry place.

General Characters.—In their fresh state the bulbs are somewhat pear-shaped or broadly ovoid ; they vary in weight from about half a pound to more than four pounds, and in size from an ordinary man's fist to that of a child's head.

The slices as seen in the pharmacies are narrow, flattish or somewhat four-sided, curved, yellowish-white or somewhat pinkish, and from about one to two inches long. They are translucent, and when moist, tough and flexible, but brittle and readily powdered when quite dry ; their taste is disagreeably bitter, mucilaginous, and acrid, but they have no marked odour.

Principal Constituents.—*Mucilage, calcium oxalate* in needle-shaped crystals (*raphides*), and two or more *bitter principles.* Until recently the only bitter principle was termed *scillitin*; but this has never been isolated. Merck has, however, of late years, isolated three distinct substances, which he has named *scillitoxin, scillipicrin,* and *scillin,* the latter of which is tasteless, but the two former have a bitter taste ; and he infers that the activity of squill as a medicine depends upon the two former, but further experiments on the action of these principles is required.

Medicinal Properties.—Expectorant and diuretic when administered in small doses ; emetic and purgative in larger ones ; and in excessive doses an irritant poison. *Dose, in powder.*—1 to 3 grains.

Official Preparations.

Acetum Scillæ, which is used in the preparation of Oxymel Scillæ ; and Syrupus Scillæ.

Pilula Ipecacuanhæ cum Scilla.
Pilula Scillæ Composita.
Tinctura Scillæ.

---

2. ALOE VULGARIS, *Lamarck.*
Barbadoes Aloe.

(Bentley and Trimen's ' Medicinal Plants,' vol. iv. plate 282.)

Habitat.—This species is a native of Northern Africa, from Morocco eastward, and probably also of peninsular India. It is likewise found on the shores of Southern Spain,

Sicily, Greece, and the Canaries; and was introduced, probably, about the beginning of the sixteenth century, into the West Indies.

Official Products and Names.—1. ALOE BARBADENSIS :— the juice, when inspissated, which flows from the transversely cut bases of the leaves of the above plant. 2. ALOIN:—a crystalline substance extracted from aloes by solvents and purified by recrystallisation.

1. Aloe Barbadensis.

Barbadoes Aloes.

Preparation and Commerce.—The active bitter juice which by inspissation forms aloes, is contained in tubular vessels, which are parallel to one another, and longitudinally arranged beneath the epidermis of the fleshy leaves of the plant, the rest of the leaf being composed of cells filled with an inert tasteless mucilaginous liquid.   Hence the best aloes of commerce is obtained in Barbadoes, by transversely cutting the fleshy leaves at their base, and collecting the juice which then flows spontaneously from the divided vessels in suitable receptacles, and subsequently evaporating it by artificial heat to a proper consistency.   It is probable, however, that in some cases it is simply inspissated by solar heat ; and it is also probable that the aloes thus prepared is more translucent than when obtained by artificial heat.

Under the head of Barbadoes Aloes the British Pharmacopœia includes all the aloes that is imported from Barbadoes and the Dutch West Indian Islands, and known in commerce as Barbadoes and Curaçao Aloes.   The Curaçao Aloes is prepared in the same manner as the ordinary Barbadoes kind.

General Characters.—Barbadoes Aloes varies in colour from deep reddish-brown or chocolate brown to dark brown or almost black ; its fracture is usually dull and waxy, or sometimes smooth and glassy, and is then termed 'Capey Barbadoes.'   It is opaque in mass, but when in thin films and held up between the eye and the light, it is translucent, and of an orange-brown tint. It is readily powdered, and the powder is of a dull olive-yellow colour. It has a strong and disagreeable odour, which is especially developed when breathed upon ; and a bitter nauseous taste.

The Curaçoa variety is commonly more glassy and translucent than the ordinary Barbadoes kind, and has a distinctive odour.

Tests.—When moistened with rectified spirit and examined in a thin stratum under the microscope, it exhibits numerous crystals. It is almost entirely soluble in proof spirit.

Principal Constituents and Medicinal Properties.—These are essentially the same as those of Socotrine Aloes, under which they are described.

Official Preparations.

Aloin (*see below*).
Enema Aloes.
Extractum Aloes Barbadensis.
Pilula Aloes Barbadensis.
Pilula Aloes et Ferri.

Pilula Cambogiæ Composita.
Pilula Colocynthidis Composita, which is also used in the preparation of Pilula Colocynthidis et Hyoscyami.

2. Aloin.

Aloin.

This is described under ' Socotrine Aloes ' (page 339).

------

3. ALOE PERRYI, *Baker.*

Socotrine Aloes.

(' Botanical Magazine,' plate 6596.)

Habitat.—A native of the Island of Socotra, and probably of the southern shores of the Red Sea and Eastern Africa.

Official Products and Names.—1. ALOE SOCOTRINA :—

the juice, when inspissated, which flows from the transversely-
cut bases of the leaves of the above plant ; and probably
other species.— 2. ALOIN :—a crystalline substance extracted
from aloes by solvents and purified by recrystallisation.

1. Aloe Socotrina.

Socotrine Aloes.

Source, Preparation, and Commerce.—The botanical
source of this aloes in the Island of Socotra, from which
its name is derived, has been definitely determined by the
investigations of Bayley Balfour, and named by Baker *Aloe
Perryi*; but it is still very doubtful, as stated in the pharmacopœia, whether this plant is its only botanical source.
Balfour says, that in the island of Socotra this kind of
aloes is obtained by allowing the juice to flow spontaneously
from the cut bases of the leaves, and then inspissating it by
sun-heat ; but the appearance of Socotrine aloes as it now
appears in the London market varies so much that its preparation in all cases cannot be uniform. Our knowledge,
indeed, of the geographical source of the official Socotrine
aloes is also by no means determined ; so that the pharmacopœia states in reference to it : ' Imported principally by
way of Bombay and Zanzibar, and known in commerce as
Socotrine and Zanzibar Aloes.'

General Characters.—The colour of socotrine aloes
(excluding the hepatic variety) is of various shades of
reddish-brown, darkening by exposure to the air ; its fracture is usually smooth and resinous-like, but sometimes
rough and irregular, and when freshly imported it is commonly soft in the interior, but after keeping it soon hardens.
In thin films when held up to the light, it is transparent
and orange ruby-red or orange-brown ; and its powder is
bright tawny reddish-brown. In other cases, Socotrine aloes
is more or less opaque and liver-coloured, and is then known
as hepatic aloes. Socotrine aloes has a strong peculiar

z

somewhat agreeable odour, which is especially developed when breathed upon ; and a very bitter taste.

Tests.—When moistened with rectified spirit and examined in a thin stratum under the microscope, it exhibits numerous crystals. It is almost entirely soluble in proof spirit.

Principal Constituents of Barbadoes and Socotrine Aloes. —The principal constituents of the official socotrine and barbadoes varieties of aloes are *volatile oil, resin,* and *aloin.* The volatile oil is in very small proportion, but to its presence the odour of aloes is due. The so-called *resin* (30 per cent.), is the substance which is precipitated when a decoction of aloes is allowed to cool ; hence it differs from ordinary resins in being soluble in boiling water. *Aloin, being official,* is described below.

There is still much difference of opinion as to the source of the medicinal properties of aloes, for while some attribute the purgative action of aloes entirely to aloin, others, while admitting this action, refer to the resin as possessing, in some degree at least, a portion of the purgative effect of the crude drug. There can be no doubt, however, that the principal purgative constituent is aloin.

Medicinal Properties of Barbadoes and Socotrine Aloes.— In small doses tonic and stomachic ; in full doses purgative and emmenagogue. *Dose.—*2 to 6 grains.

Official Preparations.

Aloin.	Pilula Aloes et Asafœtidæ.
Enema Aloes.	Pilula Aloes et Myrrhæ.
Extractum Aloes Socotrinæ, which is used in the preparation of Decoctum Aloes Compositum ; and Extractum Colocynthidis Compositum.	Pilula Aloes Socotrinæ.
	Pilula Rhei Composita.
	Tinctura Aloes.
	Tinctura Benzoini Composita.
	Vinum Aloes.

2. Aloin.

Aloin.

$$C_{16}H_{18}O_7.$$

A crystalline substance extracted from aloes by solvents and purified by recrystallisation. As obtained from the different varieties of aloes, the products differ slightly, but their medicinal properties are similar. It occurs in the proportion of from 20 to 25 per cent.

General Characters. — Usually in tufts of acicular crystals (*fig.* 47), yellow, inodorous, and having the taste of aloes. Sparingly soluble in cold water, more so in cold rectified spirit, freely

FIG. 47. Crystals of Aloin (*magnified*).

soluble in the hot fluids. Insoluble in ether. Not readily altered in acidified or neutral solutions ; rapidly altered in alkaline fluids.

Medicinal Properties.—Aloin has somewhat similar properties to aloes. *Dose.*—½ to 2 grains.

OTHER KINDS OF ALOES.

(*Not Official.*)

Besides the two official kinds of aloes of the British Pharmacopœia, other varieties are also distinguished by pharmacologists, two of which more especially are known in the London markets under the names of Cape Aloes and Natal Aloes ; the former is largely used on the Continent, but its commercial value in this country is much less than that of the official kinds. Nothing definite is known of the botanical source of Natal aloes ; or even of Cape aloes, although this has been referred to Aloe ferox, *Mill.*, and Aloe spicata, *Thunb.*

1. *Cape Aloes.*—The characteristic peculiarities of this aloes are its brilliant shining conchoidal almost vitreous fracture, hence it has been termed 'Aloe lucida'; and its peculiar somewhat sourish odour. It is also distinguished from both Socotrine and Barbadoes Aloes from not presenting any crystals when moistened with rectified spirit, and examined in a thin stratum under the microscope.

2. *Natal Aloes.*—This kind is entirely different in appearance from Cape Aloes, being greyish-brown in colour, and very opaque; hence it belongs to the hepatic kinds of aloes. Like Socotrine and Barbadoes Aloes, it is crystalline when examined in a similar way under the microscope. But the aloin it contains is different in some particulars from that derived from Barbadoes aloes (*barbaloin*); and that from Socotrine aloes (*socaloin*); hence it is termed *nataloin*. Its odour most resembles that of Cape Aloes, but is peculiar.

ORDER 2.—MELANTHACEÆ or COLCHICACEÆ.

1. VERATRUM ALBUM, *Linn.*

White Hellebore.

(Bentley and Trimen's 'Medicinal Plants,' vol. iv. plate 285.)

Habitat.—This is an Alpine and sub-Alpine species, which is common in the mountain regions of Central and Southern Europe, as Auvergne, the Pyrenees, the Alps, the Balkans, etc. It also grows throughout a great part of European and Asiatic Russia.

Part Used and Name.—VERATRI ALBI RHIZOMA :—the dried rhizome and rootlets.

(*Not Official.*)

Veratri Albi Rhizoma.

White Hellebore Rhizome.

Synonym.—Veratri Albi Radix.

Commerce and Varieties.—It is chiefly imported from Germany, and in commerce two varieties are distinguished as Swiss and Austrian. As it is generally without rootlets, it is also described as *without fibre*; or when otherwise, *with fibre*.

General Characters.—In somewhat cylindrical or conical pieces, either simple or divided above into two or more branches, each branch being commonly crowned by the concentrically arranged remains of the bases of the leaves. In length, the pieces vary from one to three or more inches; and their breadth averages about an inch at the upper end. Externally the rhizome has a greyish-brown or blackish colour, and is more or less rough from the remains of the detached rootlets, or in some cases the rhizome is covered with these rootlets; internally it presents a large central woody more or less spongy portion of a whitish or pale buff colour, separated by a fine brownish wavy-crenate ring (*nucleus sheath*) from an outer white part, which is coated by a thin dark brown or blackish portion. It is without odour, but excites sneezing when cut or bruised; its taste is bitter and very acrid.

Principal Constituents.—Recent investigations seem to prove that white hellebore rhizome contains several bases, the more important being commonly said to be *veratrine* (described under *Schœnocaulon officinale*), and *jervine*. The presence of veratrine is, however, except perhaps in very small quantity, doubtful, but an analogous base is certainly present, which has been termed *veratralbine*. (*See* Veratrum viride.)

Medicinal Properties.—Powerfully emetic and purgative when administered internally; and in excessive doses a narcotico-acrid poison. Locally applied it is sternutatory, and a powerful irritant when applied to the skin, and hence used as an insecticide. It is but little employed in this country except in veterinary medicine.

———

2. VERATRUM VIRIDE, *Soland.*
American White Hellebore.   Indian Poke.

(Bentley and Trimen's ' Medicinal Plants,' vol. iv. plate 286.)

Habitat.—This species is a marsh plant which is found growing in many parts of the North United States, Canada, and Alaska.

Official Part and Name. —VERATRI VIRIDIS RHIZOMA : the dried rhizome and rootlets.

Veratri Viridis Rhizoma.
Green Hellebore Rhizome.

Synonym.—Veratri Viridis Radix.

Collection and Commerce.—American or Green Hellebore rhizome is collected in the United States and Canada, in the autumn months ; and is then either dried in an entire state, or after being sliced in various ways.

General Characters.—Green Hellebore rhizome is either entire, or transversely or longitudinally sliced or divided, and either with or without attached rootlets ; or in some cases the rhizome and rootlets are compressed together into rectangular cakes of about an inch in thickness.   When entire it is from one to two inches or more in length, and three-quarters of an inch or more in diameter, obconical, obtuse or truncated at the apex, dark brown externally, whitish within. Frequently bearing at its upper end or base the concentrically arranged remains of leaves, and giving off on all sides numerous much shrivelled and wrinkled yellowish-white rootlets several inches long : or the latter are more or less detached and mixed with it, in which case the rhizome is marked with corresponding scars. The transverse or vertical slices exhibit a central woody portion separated from an outer part by a crenate-wavy nucleus sheath. Green hellebore is inodorous, but exciting sneezing when rubbed or powdered; its taste is bitter and very acrid.

Principal Constituents.—Very similar to those of white hellebore. Wright and Luff have described no less than six distinct bases, namely, *jervine, pseudojervine, rubijervine,* or *veratroidine,* traces of *veratrine* and *veratralbine,* and *cevadine*; all of which, except the latter, are also, according to them, present in white hellebore, but in much larger proportion.

Medicinal Properties.—Identical with, or very analogous to, white hellebore rhizome, but reputed to differ from the latter in not producing purging. Harley describes it as ' irritant and sedative like colchicum.'

Official Preparation.

Tinctura Veratri Viridis. *Dose.*—5 to 20 minims.

---

3. SCHŒNOCAULON OFFICINALE, *A. Gray.*

Sabadilla. Cevadilla.

Synonym.—ASAGRÆA OFFICINALIS, *Lindl.*

(Bentley and Trimen's ' Medicinal Plants,' vol. iv. plate 287.)

Habitat.—A native of Mexico, Guatemala, and Venezuela.

Official Parts or Products and Names.—1. SABADILLA :— the dried ripe seeds. 2. VERATRINA :—an alkaloid or mixture of alkaloids obtained from Cevadilla ; not quite pure.

1. Sabadilla.

Cevadilla.

Commerce.—The seeds were formerly imported from Vera Cruz in their pericarps ; but they are now almost entirely shipped separated from the pericarps, from La Guayra, the port of Caracas. As directed in the pharmacopœia, if imported in, or mixed with, their pericarps, these should be rejected before the seeds are used.

General Characters.—The seeds, which are *alone official*, as already stated, are one-quarter of an inch or less in length, narrow, fusiform or somewhat scimitar-shaped, prolonged above into a membranous wing, somewhat compressed, shining, wrinkled, blackish-brown. Inodorous, but when powdered producing violent sneezing; taste bitter and persistently acrid.

The pericarps when present consist of three light brown oblong pointed carpels of a papery texture ; which are united below, attached to a short stalk, and surrounded by the remains of the perianth, but separated above, and commonly open on their inner or ventral sutures.

Principal Constituents.—Cevadilla owes its properties essentially to the official alkaloid *veratrina*, which is described below. Two other alkaloids have also been described as constituents, which have been named *cevadine* and *cevadilline*.

Medicinal Properties.—Scarcely used medicinally in itself, although powerfully irritant. It is placed in the pharmacopœia simply as the source of veratrina.

Official Preparation.—Veratrina, which is described below.

2. Veratrina.

Veratrine.

Synonym.—Veratria.

An alkaloid or mixture of alkaloids obtained from Cevadilla ; not quite pure. A process by which it may be obtained is given in the British Pharmacopœia.

General Characters and Tests.—Pale grey, amorphous, without smell, but, even in the most minute quantity, powerfully irritating the nostrils ; strongly and persistently bitter, and highly acrid ; insoluble in water, soluble in spirit, in ether, and in diluted acids, leaving traces of an insoluble brown resinoid matter. It dissolves in nitric acid, yielding a yellow solution, and in sulphuric acid forming a deep red solution which exhibits a green fluorescence by reflected light. Warmed with hydrochloric acid, it dissolves with production of a blood-red colour. Heated with access of air, it melts into a yellow liquid, and at length burns away, leaving no residue.

Medicinal Properties.—It is scarcely ever employed internally in this country ; but frequently as an external remedy in the form of the official ointment, in rheumatism, neuralgia, and other painful affections, and for the destruction of pediculi. It is a powerful poison.

Official Preparation.—Unguentum Veratrinæ.

---

4. COLCHICUM AUTUMNALE, *Linn.*
Meadow Saffron.

(Bentley and Trimen's ' Medicinal Plants,' vol. iv. plate 288.)

Habitat.—This plant is plentiful in many parts of England and Ireland, but is an introduced species in Scotland. It is found growing through Middle and Southern Europe to the Mediterranean, Greece, Turkey, and the Crimea. It grows also in Northern Africa.

Official Parts and Names.—1. COLCHICI CORMUS :— the fresh corm collected about the end of June or beginning of July ; and the same stripped of its coats, sliced transversely, and dried at a temperature not exceeding

150° F. (65·5° C.).   2. COLCHICI SEMINA :—the seeds, collected when fully ripe, which is commonly about the end of July or beginning of August; and carefully dried.

1. Colchici Cormus.

Colchicum Corm.

Growth, Collection, Preservation, and Commerce.—The new corm makes its first appearance on the side of the old corm at its lower end, about the end of June or beginning of July ; this new corm flowers in the succeeding autumn, and produces its leaves in the following spring, and its seeds usually ripen about the end of July in the same year.   It then begins to shrivel, becomes leathery, and finally disappears entirely in the succeeding spring or summer.   The activity of the corm will, therefore, necessarily vary much according to its age.   It is commonly considered to be in the most active state for medicinal use when about a year old—that is, about the end of June or beginning of July, between the withering of the leaves and the sprouting forth of the flower of the new corm, and at this period it is therefore directed to be collected in the British Pharmacopœia.

The corms are usually derived from Gloucestershire, but they are also obtained from Hampshire and Oxfordshire ; and some likewise from Germany.

When not used in a fresh state they should be stripped of their coats, cut into transverse slices, and dried as directed in the British Pharmacopœia at a moderate heat.

General Characters.—The fresh corm is about one inch and a half long and an inch broad at its lower end, somewhat conical in form, slightly flattened on one side where a new corm is in process of development, and rounded on the other.   It is covered with an outer thin brown membranous coat, and an inner reddish-yellow one ; internally it is white, firm, homogeneous, and solid, and when cut yielding a milky juice of a bitter taste and disagreeable odour.

The dried slices are about one-eighth or one-tenth of an inch thick, yellowish at their circumference, moderately indented on one side, and convex on the other, so that when entire they are somewhat reniform in outline (*fig.* 48). The surfaces are firm, whitish, amylaceous; break readily with a short fracture; and have a bitter taste, but no odour.

Principal Constituents, Medicinal Properties, etc.—These are the same as colchicum seeds, and are described below, as also their official preparations.

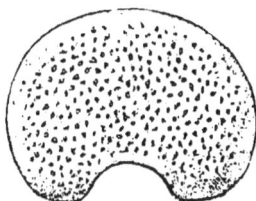

Fig. 48.—A dried transverse slice of Colchicum Corm, as seen in the pharmacies.

Substitution.—Tulip bulbs, which resemble them in size and appearance, have been sometimes substituted for Colchicum corms; but are at once distinguished by making a transverse section, when they are seen to be composed of concentrically arranged scales, whereas colchicum corms are homogeneous and solid under the same circumstances.

2. Colchici Semina.

Colchicum Seeds.

Collection.—Colchicum seeds, as directed, should be collected when fully ripe, which is commonly about the end of July or beginning of August, although sometimes earlier; after which they should be very carefully dried.

General Characters.—Sub-globular, about one-tenth of an inch in diameter, slightly pointed at the hilum, somewhat rough externally, and with a reddish-brown testa which becomes darker by keeping, or when damp. They are very hard and difficult to powder, and when broken do not present an oily appearance. They are inodorous, but have a bitter acrid taste.

Principal Constituents of Colchicum Corm and Seeds.—The principal constituent of both the corm and seeds is the

alkaloid *colchicine,* to which they owe their medicinal properties.   The corm also contains about 10 per cent. of
starch, and some odorous principle, which is lost in the
process of drying.   The properties of colchicine require
further investigation, for it has been variously described as
an alkaloid, a neutral substance, and as a weak acid.   As
usually seen it is a pale yellowish amorphous substance,
with an intensely bitter taste ; but it may be obtained in
a crystalline state.   It is a very powerful poison.

Medicinal Properties of Colchicum Corm and Seeds.—
Colchicum corm and seeds have the same properties, the
difference between them being in degree only ; both are
cathartic, emetic, diuretic, diaphoretic, cholagogue, and
sedative ; and in excessive doses act as a powerful acro-
narcotic poison.   Colchicum is generally regarded as a
specific for gout.

The dose of the powdered corm is from 2 to 8 grains ;
that of the powdered seeds is also from 2 to 8 grains.
The latter are sometimes regarded as more uniform in their
strength, and therefore more trustworthy.

Official Preparations.

1. Of COLCHICI CORMUS :—

Extractum Colchici (from the juice of fresh corms).
Dose.—½ to 2 grains.

Extractum Colchici Aceticum (from the juice of fresh
corms).   *Dose.*—½ to 2 grains.

Vinum Colchici (from the dried powdered corms).   *Dose.*
—10 to 30 minims.

2. Of COLCHICI SEMINA :—

Tinctura Colchici Seminum.   *Dose* —10 to 30 minims.

ORDER 3.—SMILACEÆ.

SMILAX OFFICINALIS, *Kunth.*

Jamaica Sarsaparilla.

(Bentley and Trimen's ' Medicinal Plants,' vol. iv. plate 289.)

Habitat.—This species is a native of Bajorque, on the river Magdalena, in New Granada, where it was first discovered by Humboldt. It has also been found in other parts of New Granada, and on the Chiriqui mountains in Costa Rica (*see* Collection, page 352). The same plant has been cultivated in Jamaica, but the root is far more starchy than that known in commerce as Jamaica Sarsaparilla ; and externally it has a light cinnamon-brown colour. It does not appear to be used in this country.

Official Part and Name.—SARSÆ RADIX :—the dried root. It is commonly known as Jamaica Sarsaparilla from having been formerly obtained from Central America by way of that island.

Before describing Jamaica sarsaparilla we must allude generally to the different commercial varieties.

Collection and Preparation.—After the removal of the earth which covers the horizontally creeping roots, the latter are cut off near the rhizome, or the latter is collected with the attached roots; they are then dried, prepared, and packed in bundles for exportation.

General Characters of Sarsaparilla Roots.—The roots of the different commercial kinds of sarsaparilla, as imported in bundles, are either without attached rhizome (*chump*), or have a portion or the whole of the rhizome attached, and sometimes also the remains of one or more of the aerial stems. The roots vary in thickness, but commonly average about that of a common writing quill, and are usually several feet in length ; the thin and shrivelled roots are technically

described as *lean*, while those which are plump and thick
are said to be *gouty*.   Frequently the main roots give off
from their sides, more or less abundantly, branched rootlets,
when they are said to be *bearded*.   Their colour varies much,
being of various shades of red, brown, or yellow.   They are
inodorous ; and in some kinds the taste is simply earthy
and mucilaginous, but in others, after chewing, slightly
acrid and feebly bitter.   A transverse section (*figs.* 49 and
50) exhibits a central cellular axis, *f,* which is surrounded
by a woody ring, *e,* and the whole is enclosed by a rind or
cortical portion, *a-d.*

FIG. 49.   Magnified view of one-half of
the transverse section of a Mealy (*Honduras*) Sarsaparilla.   *b.*   Outer cortical portion, covered by epidermis, *a.*
c, d. Inner or mealy cortical portion.
e. Woody ring.   *f.* Central cellular
axis, commonly called the pith.

FIG. 50.   Magnified view of one-half
of the transverse section of a Non-
Mealy (*Jamaica*) Sarsaparilla.   *b.*
Outer cortical portion, covered by epidermis, *a.* *c, d.* Inner cortical portion.
e. Woody ring.   *f.* Central cellular
axis, commonly called the pith.

Kinds of Sarsaparilla.—Several kinds are known in
commerce under the names of Honduras, Brazilian, Lisbon,
Guatemala, Jamaica, Mexican, Guayaquil, etc.   These may
be arranged in two divisions, as first suggested by Pereira,
called *Mealy Sarsaparillas* and *Non-Mealy Sarsaparillas.*

   1. *Mealy Sarsaparillas* (*fig.* 49).—Under this head
we include those sarsaparillas in which starch is a very
evident constituent, and under it we place the kinds known

as *Honduras*, *Brazilian*, and *Guatemala*. The characters
of mealy sarsaparillas are as follows :—The cortical layers
(*fig.* 49, *a-d*), are thick, pale-coloured, and contain a large
proportion of starch in their interior, in some cases to such
an extent that when the roots are broken it is thrown out
as a shower of whitish dust ; they vary in thickness according
to the amount of contained starch, and when very thick the
roots have a swollen appearance and are cracked transversely.
A decoction, when cool, becomes dark-blue on the addition
of a solution of iodine ; and an aqueous extract when rubbed
down with water in a mortar does not completely dissolve,
but yields a more or less turbid liquid, which becomes blue
on the addition of iodine. If sulphuric acid be added to a
transverse section, the cortical portion is but little altered in
colour, while the woody ring becomes dark purplish or
almost black.

2. *Non-Mealy Sarsaparillas.*—Under this head we include the kinds known as *Jamaica*, *Guayaquil*, and *Mexican* or *lean Vera Cruz*, the characters of which are as follows :—They are covered by a comparatively thin cortical
portion, (*fig.* 50, *a–d*), which is either non-mealy or contains but a very small proportion of starch, deep red or
brown externally, and is never cracked or swollen in appearance. A cold decoction does not yield any very evident
blue colour when a solution of iodine is added to it ; and
an aqueous extract, when rubbed down with water, is completely dissolved, forming a clear solution, which does not
become blue on the addition of a solution of iodine. If
sulphuric acid be added to a transverse section it becomes
dark red or purplish throughout.

Such are the general characters of the different kinds
of commercial sarsaparillas ; our future remarks will refer
especially to Jamaica sarsaparilla, which is *alone official.*

Sarsæ Radix.

Jamaica Sarsaparilla.

Collection.—It is said to be collected on a mountainous range in the Isthmus of Panama, adjoining the republic of Costa Rica, known as the Cordillera of Chiriqui, and then brought down to the Atlantic coast and shipped from Boca del Toro.

General Characters.—The roots are six feet or more in length, entirely free from their rhizomes, usually bent or folded and loosely packed together into bundles (*fig.* 51) of about eighteen inches in length, and from four to five

FIG. 51.   Bundle of *Jamaica Sarsaparilla.*   (Reduced.)

inches in diameter, the whole being bound together by a long root of the same drug. The roots are more or less furrowed, somewhat shrivelled, varying in thickness, but not exceeding that of an ordinary goose-quill, of a greyish-brown to deep reddish-brown colour, and have on their sides numerous branched rootlets. They have no odour ; but their taste is mucilaginous, and, when chewed, feebly bitter, and faintly acrid.

Their relation to tests, etc., is the same as already described when speaking of non-mealy sarsaparillas, of which Jamaica sarsaparilla may be taken as a type. Jamaica sarsaparilla is sometimes termed *red sarsaparilla* or *red bearded sarsaparilla.* It possesses all the characters which in Britain are regarded as those of the best sarsaparillas —that is, a deep reddish-brown colour, a great number of branched rootlets, and an acrid taste : it also yields a large amount of extract.

Principal Constituents.—The principal constituents of the different sarsaparillas are *starch, calcium oxalate crystals* (*raphides*), *resin,* a small amount of *volatile oil,* and a peculiar principle named *smilacin* or *parillin.* The latter is commonly regarded as the active principle, but this has not been in any degree established. It is said to be a glucoside closely resembling, if not identical with, saponin. It is coloured reddish-brown by the action of concentrated sulphuric acid; hence, if a drop of the latter, as already noticed under Non-mealy Sarsaparillas, be applied to a transverse section of the root of Jamaica sarsaparilla, both the cortical portion and woody ring become dark-red or purplish.

Medicinal Properties.—It is usually regarded as alterative, tonic, diaphoretic, and diuretic; but many believe it to be almost, if not entirely, inert.

Official Preparations.

Decoctum Sarsæ. | Decoctum Sarsæ Compositum.
Extractum Sarsæ Liquidum.

SUB-CLASS II.—GLUMACEÆ.

ORDER GRAMINACEÆ.

1. TRITICUM SATIVUM, *Lam.*

Wheat.

Synonym.—TRITICUM VULGARE, *Villars.*

(Bentley and Trimen's ' Medicinal Plants,' vol. iv. plate 294.)

Habitat.—The cultivation of this plant is coeval with the history of man; it is essentially the cereal of temperate climates. No form of it has ever been seen wild, nor, indeed, any species very closely resembling it.

Official Products and Names.—1. FARINA TRITICI :—the grain, ground and sifted. 2. AMYLUM :—the starch

procured from the grains of common wheat, Triticum
sativum, *Lam.* ; maize, Zea Mays, *Linn.* ; and rice, Oryza
sativa, *Linn.*   3.   MICA PANIS :—the soft part of bread made
with wheaten flour.

1. Farina Tritici.
Wheaten Flour.

Preparation.—The grains of wheat are commonly termed
seeds, but properly speaking they are a kind of fruit called a
caryopsis.   As seen in the markets they have been divested
of their pales (*chaff*) ; they are prepared for use by grinding
and sifting, by which the farina or flour is separated from the
integuments or bran ; the latter forming from 25 to 33 per
cent.

General Characters and Principal Constituents.—Wheaten
flour is white, colourless, and nearly tasteless.   The principal
constituents are *starch, gluten, albumen, dextrine,* and *sugar,*
the proportion of which varies much in different grains, but
their characters do not come within our province.

Medicinal Properties.—Wheaten flour can scarcely be regarded as a medicinal substance, its use being confined to
sprinkling over the surface in burns and scalds ; and on the
skin in various itching and burning eruptions, as nettle-rash,
and in erysipelatous inflammations.

Official Preparation.—Cataplasma Fermenti.

2. Amylum.
Starch.

Source.—As already noticed, starch may now be officially obtained from either the grains of wheat, maize, or
rice.

Characters and Tests.—In fine powder, or in irregular
angular or columnar masses, which are readily reduced to
powder ; white, inodorous.   When lightly rubbed in a
mortar with a little cold distilled water, the mixture is
neither acid nor alkaline to test-paper, and the filtered

liquid does not become blue on the addition of solution of
iodine. Mixed with boiling water and cooled, it gives a
deep blue colour with iodine. Under the microscope these
varieties of starch present the following characters :—

1. *Wheat starch* : A mixture of large and small
granules, which are lenticular in form, and marked with
faint concentric striæ surrounding a nearly central hilum
(*fig.* 52). 2. *Maize starch* : Granules more uniform in size,
somewhat smaller than the large granules of wheat starch,

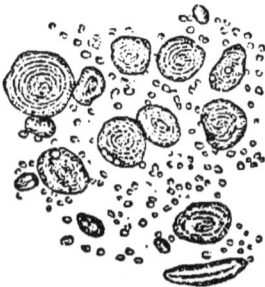

FIG. 52.—*Wheat Starch* (magnified
250 diameters).

FIG. 53.—*Maize Starch* (magnified
250 diameters).

frequently polygonal, and having a very distinct hilum but
without evident concentric striæ (*fig.* 53). 3. *Rice starch* :
Granules extremely minute, nearly
uniform in size, polygonal, hilum
small, and without striæ (*fig.* 54).

Medicinal Properties.—Used for
sprinkling over inflamed surfaces, etc.
and in the form of mucilage for stif-
fening bandages for fractured limbs
and other purposes. As an enema it
has demulcent properties, when used
in the form of the official mucilage,
and is also useful in the same form

FIG. 54.—*Rice Starch* (magnified 250 diameters).

as a vehicle for the preparation of more active enemas.
Starch may be also employed as an antidote to poisoning
by iodine, etc.

A A 2

Official Preparations.

Glycerinum Amyli, which is also used in the preparation of Suppositoria Acidi Carbolici cum Sapone; Suppositoria Acidi Tannici cum Sapone; Suppositoria Morphinæ cum Sapone.

Mucilago Amyli, which is also used in the preparation of Enema Aloes; Enema Magnesii Sulphatis; Enema Opii; Enema Terebinthinæ.

Pulvis Tragacanthæ Compositus.

Suppositoria Acidi Tannici cum Sapone.

Suppositoria Morphinæ cum Sapone.

3. Mica Panis.

Crumb of Bread.

The official crumb of bread is the soft part of bread made with wheaten flour.

Medicinal Properties.—In the form of a poultice it is a valuable emollient application, and as a vehicle for other more active substances, etc. It is also used as a pill excipient.

Official Preparation.—Cataplasma Carbonis.

---

2. ZEA MAYS, *Linn.*

Maize. Indian Corn.

(Bentley and Trimen's 'Medicinal Plants,' vol. iv. plate 296.)

Habitat.—Not known in the wild state.

Official Product and Name.—The grains (*caryopsides*) are official as one of the sources of Amylum (*see* Amylum).

3. ORYZA SATIVA, *Linn.*

Rice.

(Bentley and Trimen's ' Medicinal Plants,' vol. iv. plate 291.)

Habitat.—An undoubted native of India, where its wild form is common ; probably also of China.   The grains from cultivated plants are, however, alone in use.

Official Product and Name.—The grains (*caryopsides*) are official as one of the sources of Amylum (*see* Amylum).

4. HORDEUM DISTICHON, *Linn.*

Barley.

Synonym.—HORDEUM VULGARE, *Linn.*

(Bentley and Trimen's ' Medicinal Plants,' vol. iv. plate 293.)

Habitat.—It is not now known in a wild state, though some closely allied species are found growing wild.   Its native country is supposed to have been the Southern Caucasus and the shores of the Caspian Sea.

Official Part and Name.—HORDEUM DECORTICATUM : the dried seed ; divested of its integuments.

Hordeum Decorticatum.

Pearl Barley.

Cultivation.—The seed is directed to be obtained from plants cultivated in Britain.

General Characters.—When the grains or caryopsides are deprived of their pales, with which they are commonly enclosed as seen in commerce, they form *Scotch, Hulled,* or *Pot Barley* ; and when all the integuments are removed and the seeds rounded and polished, they constitute the *official* or *Pearl Barley.*   This is rounded, white, with a trace of the longitudinal furrow of the grain, in which are the remains of the yellowish-brown integuments. It has the ordinary farinaceous taste and odour of the cereal grains generally.

Principal Constituents. — Essentially the same as the grains of wheat, but the nitrogenous principles are less in proportion.

Medicinal Properties.—In the form of the official decoction, pearl barley forms a demulcent and nutritious drink ; and a vehicle for the administration of other more active medicines.

Official Preparation.—Decoctum Hordei.

———— ————

5. SACCHARUM OFFICINARUM.
Sugar Cane.

(Bentley and Trimen's ' Medicinal Plants,' vol. iv. plate 298.)

Habitat.—The wild form of this plant is not known with any degree of certainty, but without doubt originally a native of India.

Official Products and Names.—1. SACCHARUM PURIFICATUM :—refined sugar. 2. THERIACA :—the uncrystallised residue of the refining of sugar.

1. Saccharum Purificatum.
Refined Sugar.

Synonym.—Sucrose.

$$C_{12}H_{22}O_{11}.$$

Botanical Source.—It will be noticed that in the present pharmacopœia, refined sugar or sucrose is alone mentioned without giving any botanical source, hence any sucrose may be used, if refined, from whatever plant derived, and not necessarily, therefore, from the sugar cane as directed in all preceding pharmacopœias.

General Characters and Tests. — Compact crystalline conical loaves, known in commerce as lump sugar. Readily and completely soluble in water, forming a clear bright syrup which yields no red or yellowish precipitate, or scarcely a trace, on heating it to near the boiling point of water for a short time with a little solution of sulphate of copper and excess of solution of potash.

Medicinal Properties.—In itself refined sugar is of little importance in a medicinal point of view, but it is largely used in pharmacy as a preservative agent, to give flavour and consistence, and for various other purposes.

Official Preparations.

Confectio Rosæ Caninæ.
Confectio Rosæ Gallicæ.
Confectio Sennæ.
Extractum Sarsæ Liquidum.
Ferri Carbonas Saccharata.
Liquor Calcis Saccharatus.
Mistura Ferri Composita.
Mistura Guaiaci.
Mistura Spiritus Vini Gallici.
Pilula Ferri Iodidi.

Pulvis Amygdalæ Compositus.
Pulvis Cretæ Aromaticus.
Pulvis Glycyrrhizæ Compositus.
Pulvis Tragacanthæ Compositus.
Sodii Citro-tartras Effervescens.
All the Syrups and Lozenges.

Confectio Rosæ Gallicæ ; Ferri Carbonas Saccharata ; Pulvis Amygdalæ Compositus; Syrupus; and Syrupus Aurantii, are also contained in other official preparations.

2. Theriaca.

Treacle.

General Characters.—A thick fermentable syrup of a golden colour, very sweet ; not crystallising by rest or spontaneous evaporation. Specific gravity about 1·40.

Test.—Free from empyreumatic odour or flavour.

Medicinal Properties.—Slightly laxative. It is much used

in pharmacy to give consistency to pill masses, as a preservative agent, and for other purposes.

Official Preparations.

Pilula Aloes et Myrrhæ.

Pilula Asafœtidæ Composita.

Pilula Conii Composita.

Pilula Ipecacuanhæ cum Scilla.

Pilula Rhei Composita.

Pilula Scillæ Composita.

Tinctura Chloroformi et Morphinæ.

DIVISION II. *GYMNOSPERMIA.*

ORDER CONIFERÆ or PINACEÆ.

1. JUNIPERUS COMMUNIS, *Linn.*

Juniper.

(Bentley and Trimen's ' Medicinal Plants,' vol. iv. plate 255.)

Habitat.—This plant, under one or other of its varieties, has a very extensive distribution, extending throughout Europe and North Africa, Asia northwards from the Himalayas, Japan, and North America ; the dwarf form reaches far into the Arctic regions, occurring in Greenland and Kamtschatka. It is widely diffused in England, growing in hilly places.

Official Product and Name.—OLEUM JUNIPERI :—the oil distilled in Britain from the full-grown unripe green fruit.

General Characters of the Fruit.—The fruit is commonly termed a berry, although properly a *galbulus* (*fig.* 55). In the unripe state it is green, and about one third of an inch or the size of a small pea or black currant; but when ripe, which is not till the second year, it is purplish-black, and covered with a whitish-blue bloom. It is marked at the summit (*fig.* 55, *b*, *c*) by three radiating furrows, caused by the three fleshy bracts of which it is composed (*fig.* 55, *a*) having completely united, except at their tips ; and each fruit contains three bony seeds enveloped in pulp. At the

FIG. 55.—Fruit of *Juniperus communis.* *a.* Fruit approaching maturity. *b.* Mature fruit. *c.* Ripe fruit, seen from the apex. *d.* Ripe fruit, seen from the base.

base of the fruit are the empty minute bracts or scales of the cone of the female flower, arranged in a stellate manner (*fig.* 55, *d*). When bruised the odour is agreeably aromatic ; and the taste sweetish, somewhat spicy, and terebinthinous.

Oleum Juniperi.

Oil of Juniper.

Preparation.—The official oil of juniper, to which the properties of the fruits are due, may be readily obtained by distilling them with water, the amount varying from about 0·5 to 1·2 per cent.

General Characters.—Colourless or pale greenish-yellow, neutral, limpid, transparent, levogyrate, specific gravity about 0·70 to 0·90, with the characteristic odour of the fruit, and a warm aromatic taste.

Medicinal Properties.—Juniper fruit and oil are carminative, stimulant, and more especially diuretic. *Dose of Oil.*— 1 to 4 minims.

Official Preparation of Oleum Juniperi.

Spiritus Juniperi. *Dose.*—$\frac{1}{2}$ to 1 fluid drachm. Spirit of Juniper is also an ingredient in Mistura Creasoti.

362   *Sabinæ Cacumina.*   [*Gymnospermia*]

2. JUNIPERUS SABINA, *Linn.*

Savin.

(Bentley and Trimen's ' Medicinal Plants,' vol. iv. plate 254.)

Habitat.—This species is widely distributed through the Northern Hemisphere, occurring in Central and Southern Europe, but not extending to the British Islands. In Asia it is found in the Caucasus, the Caspian districts, the Altai Mountains, and the whole of Siberia. It is also found in Canada, Newfoundland, and the Northern United States.

Official Parts or Products and Names.—1. SABINÆ CACUMINA :—the fresh and dried tops. 2. OLEUM SABINÆ : the oil distilled in Britain from the fresh tops.

1. Sabinæ Cacumina.

Savin Tops.

Collection.—They are directed to be collected in spring, from plants cultivated in Britain, and should consist of the young green tender twigs stripped from the more woody portions of the plants.

General Characters.—These twigs are densely covered with minute imbricated adpressed rhomboid dark green (or when dried yellowish-green) leaves, with a large oval depressed central oil gland on their back. When rubbed or bruised (more especially when fresh), they have a strong peculiar odour ; and an acrid bitter disagreeable taste.

Principal Constituents. — The essential constituent is the official *volatile oil* described below.

Medicinal Properties.—When given internally, or locally applied, it is an irritant : its effects being more especially manifested on the uterus, and it is, therefore, a powerful emmenagogue. In excessive doses it is an energetic poison ;

and from being a powerful abortifacient its use is contra-indicated in pregnancy. *Dose, in powder.*—4 to 10 grains.

Official Preparations.—Oleum Sabinæ, from fresh tops. *Dose.*—1 to 4 minims. Tinctura Sabinæ, from the dried and coarsely powdered tops. *Dose.*—20 minims to 1 fluid drachm. Unguentum Sabinæ, from fresh tops.

2. Oleum Sabinæ.

Oil of Savin.

Preparation.—Oil of savin is distilled from the fresh tops, which yield from 2 to 4 per cent.

General Characters.—Colourless or pale yellow, with the odour of the plant, and a bitter acrid taste. It is dextro-gyrate, neutral, specific gravity about 0·915, and freely soluble in absolute alcohol.

Medicinal Properties.—Similar to the tops ; and more reliable and convenient for internal use.

3. PINUS PINASTER, *Solander.*

Maritime Pine. Cluster Pine.

(Bentley and Trimen's ' Medicinal Plants,' vol. iv. plate 256.)

Habitat.—A native of South-Western Europe, in the neighbourhood of the sea-coast in Spain and Portugal, Southern and Western France ; it extends eastward in the Mediterranean basin to Algeria, Corsica, Southern Italy, Sicily, and Greece.

Official Products and Names.—1. OLEUM TEREBIN-THINÆ :—the oil distilled, usually by aid of steam, from the oleo-resin (turpentine) obtained from this and the other species of Pinus, as described under ' Pinus australis, *Mich.*' 2. RESINA :—the residue left after the distillation of the oil of turpentine from the crude oleo-resin (turpentine) of various species of Pinus, *Linn.* This is also described under ' Pinus australis.'

This species also yields *tar* (Pix Liquida), resembling that from Pinus sylvestris, *Linn.*, under which plant it is described ; and two *non-official substances*, namely, *Pitch*, described under ' Pinus sylvestris, *Linn.*,' and *Galipot.* The latter term is applied in France to the turpentine which concretes on the trunk of this tree. It corresponds to the official *Thus Americanum*, which is described under ' Pinus australis, *Mich.*'

---

4. PINUS SYLVESTRIS, *Linn.*
Scotch Fir.

(Bentley and Trimen's ' Medicinal Plants,' vol. iv. plate **257.**)

Habitat.—It forms vast woods in many parts of Northern Europe and Asia, extending into the Arctic regions ; it is also a native of the Central European chains, and extends into the Caucasus, Armenia, and Cappadocia. In Britain, although extensively planted, it is only truly native in a few of the Highland forests of Scotland.

Official Products and Names.—1. OLEUM TEREBINTHINÆ. 2. RESINA. Both these are described under ' Pinus australis.' 3. PIX LIQUIDA :—a bituminous liquid, obtained from the wood of this and other species of Pinus, by destructive distillation. 4. OLEUM PINI SYLVESTRIS :—the oil distilled from the fresh leaves.

1. OLEUM TEREBINTHINÆ, and 2. RESINA, are described under ' Pinus australis.'

3. Pix Liquida.
Tar.

Synonym.—Liquid Pitch.

Preparation and Commerce.—Tar is obtained by the destructive distillation of fir-wood in the northern parts of

Europe and in America ; the former, which is generally regarded as the best, is known in commerce as *Archangel Tar* or *Stockholm Tar*, and is very largely procured from *Pinus sylvestris* ; and the latter, termed *American Tar*, is principally derived from *Pinus australis.*

General Characters.—A dark-brown or blackish semi-liquid substance, of a well-known peculiar aromatic odour. Water agitated with it acquires a pale yellowish-brown colour, sharp empyreumatic taste, and acid reaction.

Principal Constituents.—Tar is a very complex substance, but it consists principally of various *liquid hydrocarbons, empyreumatic resin, acetic acid, creasote,* and *pyrocatechin.*

Medicinal Properties.—Tar, both externally applied, and given internally, acts as a stimulant ; it is used externally in skin diseases, etc., and internally in the form of vapour from heated tar in pulmonary affections, etc.

Official Preparation.—Unguentum Picis Liquidæ.

PITCH or BLACK PITCH, which was *formerly official,* is the residuum left after the distillation of tar. At ordinary temperatures it is an opaque black solid substance, which breaks with a shining conchoidal fracture. It has little taste, but a disagreeable odour.

It consists of *resin,* combined with various other em-pyreumatic resinous substances which are commonly known under the name of *pyretin.*

4. Oleum Pini Sylvestris.

Fir-wool Oil.

Synonym.—Oil of Scotch Pine Leaf.

Preparation and Commerce.—It is prepared in Germany by distillation from the fresh leaves.

General Characters and Tests.—Colourless or nearly so, with an agreeable aromatic odour of the fresh pine and somewhat resembling that of lavender, and a pungent but not unpleasant taste. Its specific gravity should not be below 0·870. It is soluble in about seven times its volume of rectified spirit.

Medicinal Properties.—A mild and useful stimulant when used externally, and as an inhalation.

Official Preparation.—Vapor Olei Pini Sylvestris.

5. PINUS AUSTRALIS, *Michaux.*
Broom Pine.   Pitch Pine.
Synonym.—PINUS PALUSTRIS, *Mill.*

(Bentley and Trimen's ' Medicinal Plants,' vol. iv. plate 258.)

Habitat.—It forms vast woods in Virginia, Carolina, Georgia, and Florida; it is not found in the Northern United States.

Official Products and Names.—1. OLEUM TEREBINTHINÆ :—the oil distilled, usually by aid of steam, from the oleo-resin (turpentine) obtained from Pinus australis, *Mich.*, Pinus Tæda, *Linn.*, and sometimes from Pinus Pinaster, *Solander*, and Pinus sylvestris, *Linn.* ; rectified if necessary. 2. RESINA :—the residue left after the distillation of the oil of turpentine from the crude oleo-resin (turpentine) of various species of Pinus, *Linn.* 3. THUS AMERICANUM:—the concrete turpentine which is scraped off the trunks of Pinus australis, *Mich.*, and Pinus Tæda, *Linn.*

Nature of Turpentine, Extraction, Varieties, and Commerce.—Turpentine is an oleo-resin which is secreted by several coniferous plants besides those mentioned above as yielding the official turpentine. It is contained in what have been called *resin-ducts* or *turpentine vessels*, which are a kind of *receptacle of secretion.* These receptacles are situated in the bark or the wood, or in both bark and wood, and the turpentine either exudes spontaneously; or more commonly it is obtained by incising the trees in different ways. Thus, in the United States, cavities, which are technically called *boxes*, are cut in the alburnum, and the oleo-resinous juice, which then flows into them, is subsequently collected, and in this state it constitutes the *common* or *crude turpentine* of commerce. The Bordeaux or French turpentine is also collected by placing suitable vessels below incisions. Some of the turpentine which concretes spontaneously upon the trunk is also scraped off and kept separately, and is known in the market as *scrape*; and it is this as obtained from *Pinus australis* and *Pinus Tæda*, which constitutes our official *Thus Americanum.*

This crude turpentine or oleo-resin is at first fluid, but becomes gradually thicker by age and exposure to the air, a change which is due in part to the volatilisation, and in part probably to the resinification, of its volatile oil. There are three varieties of the crude oleo-resin or common turpentine, known in commerce, namely, American Turpentine, Bordeaux or French Turpentine, and Russian Turpentine, all of which are official in the British Pharmacopœia as sources of the Oil of Turpentine of that volume. The first kind is essentially obtained from *Pinus australis*, but, to some extent, also from *P. Tæda*; the second is from *P. Pinaster*; and the third from *P. sylvestris*. But by far the larger proportion of turpentine used in Britain is the American kind. It is principally extracted in Virginia and the Carolinas, and is imported chiefly by way of Boston. Bordeaux Turpentine, as its name implies, is imported from Bordeaux, in the neighbourhood of which city it is largely obtained; and Russian Turpentine is procured in Russia and Finland.

The amount of volatile oil varies in different specimens of crude turpentine from about 15 to 18 per cent., Bordeaux or French turpentine yielding more than the American kind.

General Characters of Crude Turpentine.—As seen in

commerce, American turpentine is a yellowish-white viscid liquid, which is somewhat opaque at first, but by exposure to the air it becomes transparent ; its odour is aromatic and agreeable, and its taste warm and bitterish. In cold weather it becomes a soft solid. When imported it is almost always contaminated with leaves, chips, and other impurities, from which it is separated by melting and straining, and is then frequently termed *refined turpentine.*

Bordeaux and Russian turpentines have essentially the same characters as American turpentine, but distinctive odours and tastes.

1. Oleum Terebinthinæ.

Oil of Turpentine.

This volatile oil is commonly, though incorrectly, called *spirits* or *essence of turpentine.*

Preparation.—This *official oil* is distilled from the oleo-resin described above, and usually by aid of steam. It is either distilled in this country or imported, and then princi-pally from North Carolina.

General Characters and Constituents. — It is a limpid, colourless, very inflammable liquid, with a strong peculiar odour, which varies in the different kinds, and a pungent and bitterish taste. It commences to boil at about 320° F. (160° C.), and almost entirely distils below 356° F. (180° C.), little or no residue remaining. Its specific gravity varies from 0·856 to 0·870 ; and while American oil of turpentine is dextrogyre, that of the Bordeaux or French kind is levogyre. It is insoluble in water, but readily solu-ble in absolute alcohol and ether. It is a mixture of various hydrocarbons, all of which have the formula $C_{10}H_{16}$; and probably also of a small amount of some oxygenated oils, but not one of these has been isolated.

Medicinal Properties. In small doses it is stimulant, diuretic, antispasmodic, and astringent ; its action being more especially marked on the mucous membranes of the genito-urinary organs.   In large doses it is purgative and anthelmintic.   When applied locally it is rubefacient and counter-irritant.

Dose.—10 minims to 4 fluid drachms.

Official Preparations.

Confectio Terebinthinæ.	Linimentum Terebinthinæ
Enema Terebinthinæ.	Aceticum.
Linimentum Terebinthinæ.	Unguentum Terebinthinæ.

2. Resina.

Resin.

Preparation.—As already stated, resin is the residue left after the distillation of the oil of turpentine from the crude oleo-resin (turpentine).   It is also known as *colophony* and *yellow resin.*   In this state it always contains a little water; but when freed from water by fusion and long exposure to heat, it acquires a darker colour, and is then known as *black resin.*   The former is the official resin.

General Characters and Constituents.—Translucent, yellowish, compact, brittle, pulverisable.   It breaks with a shining fracture ; its odour and taste are faintly terebinthinate ; it is easily fusible, and burns with a dense yellow flame and much smoke.

It is essentially composed of *abietic anhydrid*, this forming from 80 to 90 per cent. of its substance.

Medicinal Properties.—Mildly stimulant, and much used externally in plasters, ointments, etc.

Official Preparations.

Charta Epispastica.	Emplastrum Resinæ.
Emplastrum Calefaciens.	Emplastrum Saponis.
Emplastrum Cantharidis.	Unguentum Resinæ.
Emplastrum Picis.	Unguentum Terebinthinæ.
Emplastrum Plumbi Iodidi.	

B B

Emplastrum Resinæ is also an ingredient in Emplastrum Belladonnæ; Emplastrum Calefaciens; and Emplastrum Opii. Emplastrum Saponis is also used in the preparation of Emplastrum Belladonnæ; and Emplastrum Calefaciens.

3. Thus Americanum.

Common Frankincense.

Nature.—This substance, which is also incorrectly known as *Gum Thus*, is, as previously stated, the crude turpentine that concretes spontaneously upon the surface of the pine trees during the collection of the ordinary American turpentine. It is obtained by scraping the trees, and hence its commercial name of *scrape*. It is an *oleo-resin* with a small proportion of *volatile oil.*

General Characters.—When fresh it is a softish yellow opaque tough solid, with the same odour as crude American turpentine; but by keeping it becomes dry and brittle, darker in colour, and of a milder odour. It should be freed from all the impurities with which it is commonly more or less contaminated when imported, such as chips, leaves, etc., by melting and straining.

Medicinal Properties.—Externally as a mild stimulant in the form of a plaster.

Official Preparation.—Emplastrum Picis.

---

6. PINUS TÆDA, *Linn.*

Loblolly.   Oldfield Pine.   Frankincense Pine.

(Bentley and Trimen's ' Medicinal Plants,' vol. iv. plate 259.)

Habitat.—It is a native of the Southern United States.

Official Products and Names.—1. OLEUM TEREBINTHINÆ.   2. RESINA.   3. THUS AMERICANUM.

These substances have been fully described under ' Pinus australis,' which is their principal source.

7. PINUS LARIX.

Larch.

Synonym.—ABIES LARIX, *Lam.*

(Bentley and Trimen's ' Medicinal Plants,' vol. iv. plate 260.)

Habitat.—A native of mountain regions in Central Europe.

Official Part and Name.—LARICIS CORTEX :—The bark. Collected in spring, deprived of its outer rough portion, and dried.

Laricis Cortex.

Larch Bark.

Preparation.—It should be collected in the spring, and prepared for use as above directed, when it consists of the inner bark and a variable proportion of the middle and outer layers of the bark.

General Characters.—Larch bark is in flattish pieces or quills of varying lengths and sizes. The outer surface is somewhat uneven, dark red or rosy, with frequently whitish or yellowish intervening portions of liber ; the inner surface is nearly smooth, yellowish-white or pinkish-red according to its age, and separable into layers. It breaks with a close fracture, except as regards the liber, which is somewhat fibrous, and the fractured surface, except internally, is of a deep carmine-red colour. The odour is slightly balsamic and terebinthinous ; and it has an astringent taste.

Principal Constituents.—A volatile crystallisable principle, termed *larixin* or *larixinic acid, tannic acid,* and a very little *turpentine,* which is commonly known as *Venice* or *Larch turpentine.* This turpentine is a slightly turbid thick honey-like liquid, of a pale yellow or greenish-yellow colour ; an acrid bitter somewhat aromatic taste ; and with an odour

resembling common turpentine, but milder and not so agreeable. It consists of *volatile oil* and *resin*; but it thickens
very slowly by exposure to the air. There is but very little
of this turpentine in the bark; but it is obtained commercially
by boring into the heart-wood, where it is principally found.

Medicinal Properties.—Larch bark is regarded as a stimulant expectorant, astringent, and diuretic; and is supposed to have an especial effect in chronic bronchitis with
abundant secretion.

Official Preparation.—Tinctura Laricis. *Dose.*—20 to 30
minims.

---

8. PINUS PICEA, *Du Roi.*

Spruce Fir.   Norway Spruce.

Synonyms.—PINUS ABIES, *Linn.*; ABIES EXCELSA, *DC.*

(Bentley and Trimen's ' Medicinal Plants,' vol. iv. plate 261.)

Habitat.—This tree is very abundant in the North of
Europe, extending into the Arctic Circle in Lapland and
Finland. It is found on all the mountain ranges of Europe,
but is rare in the Pyrenees. Although commonly planted
here, it is not a native of the British Isles.

Official Product and Name.—PIX BURGUNDICA:—the
resinous exudation obtained from the stem; melted and
strained.

Pix Burgundica.

Burgundy Pitch.

Production and Commerce.—It is not produced in
Burgundy, as its common name would imply, but on a very
large scale in Finland, and in the Grand Duchy of Baden,
Austria, and Switzerland. It is either a spontaneous exudation, or obtained after incision, by scraping the trees, and
then subsequently purifying it by melting in hot water and
straining.

General Characters and Test.—Hard and brittle when cold, yet gradually taking the form of the vessel in which it is kept.   It is somewhat opaque, dull reddish-brown or yellowish-brown in colour, and its fracture is clear and conchoidal.   The odour is agreeable and aromatic, especially when heated; and its taste is sweet, aromatic, and without bitterness.   It is readily soluble in glacial acetic acid.

Principal Constituents.—Like that of the other crude turpentines of the Coniferæ—*volatile oil* in small proportion, and *resin.*

Adulteration and Substitution.—It is very extensively adulterated; indeed artificial products are frequently substituted for it, but these are readily known by the peculiar agreeable odour of the genuine drug being absent, and by the other characters and test already given of true Burgundy Pitch.

Medicinal Properties.—Slightly stimulant when applied to the skin in the form of a plaster.

Official Preparations.

Emplastrum Ferri. | Emplastrum Picis.

---

9. PINUS BALSAMEA, *Linn.*

Balsam Fir.   Balm of Gilead Fir.

Synonym.—ABIES BALSAMEA, *Mill.*

(Bentley and Trimen's ' Medicinal Plants,' vol. iv. plate 263.)

Habitat.—It is found in Labrador, Nova Scotia, and other parts of Canada as far north as 62° ; it extends in the Northern and Western United States, as far south as Pennsylvania, and along the mountains even to Virginia.

Official Product and Name.—TEREBINTHINA CANADENSIS :—the turpentine obtained by puncturing or incising the bark of the trunk and branches.

Terebinthina Canadensis.

Canada Turpentine.

Synonym.—Canada Balsam.

Extraction and Commerce.—It is obtained as stated in the official description, and as the oleo-resinous juice runs out it is collected in a bottle or other suitable vessel.

It is principally obtained in Lower Canada, but to some extent also in the State of Maine. It is exported from Quebec and Montreal.

General Characters.—When fresh it is somewhat turbid, but soon becomes clear and transparent, so that as seen in commerce it is a pale yellow and faintly greenish transparent oleo-resin, of the consistence of thin honey, with a peculiar agreeable aromatic terebinthinate odour, and a slightly bitter feebly acrid taste. By exposure to the air it thickens very slowly, and becomes darker in colour; and ultimately, after some time, it dries into a transparent adhesive varnish. When mixed with about one-sixth of its weight of magnesia it solidifies and becomes of a pilular consistence.

Principal Constituents.—It is an oleo-resin like the other crude turpentines; hence its common name of *Canada Balsam* is incorrect, as it contains neither benzoic nor cinnamic acids, the presence of one of which at least, as we have seen (page 135), is necessary to constitute a balsam.

Medicinal Properties.—Similar to the other turpentines, but its use as a medicinal agent is almost confined to its employment as a constituent of Blistering Paper and Flexible Collodion.

Official Preparations.

Charta Epispastica. | Collodium Flexile.

SUB-KINGDOM II.

CRYPTOGAMIA OR FLOWERLESS PLANTS.

DIVISION I. *CORMOPHYTA.*

ORDER FILICES.

ASPIDIUM FILIX-MAS, *Swartz.*

Male Fern.

Synonyms.—LASTREA FILIX-MAS, *Presl*; NEPHRODIUM
FILIX-MAS, *Michaux.*

(Bentley and Trimen's 'Medicinal Plants,' vol. iv. plate 300.)

Habitat.—It is one of our commonest ferns, and in the
form of some of its numerous varieties it has a very wide
range over the world. Thus it is found in all parts of
Europe, Temperate Asia, Northern India, North and South
Africa, the temperate parts of the United States, and the
Andes of South America.

Official Part and Name.—FILIX MAS :—the rhizome
with the persistent bases of the petioles.

Filix Mas.

Male Fern.

Collection and Preservation.—It is directed to be collected
late in the autumn, divested of its scales, roots (*fig.* 56), and
all dead portions, and carefully dried with a gentle heat.
It is also officially stated that it should not be used if more
than a year old.

General Characters.—If obtained and dried as directed,
male fern should be from three to six or more inches in

length, and the rhizome itself from three-quarters of an inch to about an inch in diameter, but, being entirely covered by the hard persistent curved angular dark-brown bases of the petioles (*fig.* 56, *b, b, b*), it is apparently two or more inches. The rhizome is brown externally, and yellowish-white or brownish internally. Its odour is feeble but disagreeable ;

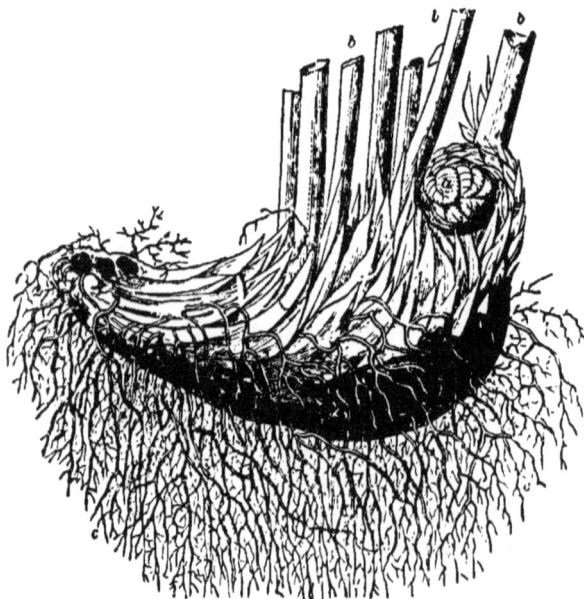

Fig. 56.— Fresh entire rhizome of *Aspidium Filix-mas. a.* Circinate young frond. *b, b, b.* Bases of the petioles. *c, c.* Roots.

and the taste sweetish and astringent at first, but subsequently bitter and nauseous.

Principal Constituents.—The principal constituents are a *green fatty oil*, a little *volatile oil, resin, tannic acid,* and *filicic acid.* The latter is crystalline, granular, and colourless, and is regarded by Buchheim as the source of the medicinal activity of the drug.

Substitution.—The rhizomes and petioles of other allied ferns are said to be sometimes substituted for male fern. The authors of ' Pharmacographia ' state that these may be readily distinguished by examining with a magnifying lens the transverse sections of their petiole-bases ; for while those of *Aspidium Filix-mas* exhibit eight fibro-vascular bundles—in those of *Asplenium Filix-fœmina,* Bernh., *Aspidium Oreopteris,* Swartz, and *Aspidium spinulosum,* Swartz, only two bundles can be observed.

Medicinal Properties.—Male fern is a valuable anthelmintic, and may be especially used in all cases of tapeworm.

Official Preparation. — Extractum Filicis Liquidum. The *dose* given in the British Pharmacopœia, is 15 to 30 minims ; but Lauder Brunton says that this is too small, and gives the dose as $1\frac{1}{2}$ to 2 fluid drachms.

Division II. *THALLOPHYTA.*

Order I.—LICHENES.

1. ROCCELLA TINCTORIA, *DC.*

Orchella Weed. Dyer's Weed.

(Bentley and Trimen's ' Medicinal Plants,' vol. iv. plate 301.)

Habitat.—This lichen is found growing on sea-shore rocks, within reach of the spray, in nearly all the warm parts of the globe.

Official Product and Name.—Litmus :—a blue pigment prepared from various species of Roccella, *DC.*

Litmus.

Preparation and Commerce.—Litmus appears to be prepared exclusively in Holland, and the process consists in macerating the above species of Roccella, as well as several other lichens in a coarsely ground state, in a mixture of urine, with lime, potash, or soda for several weeks, by which a kind of fermentation is set up, and the mass first becomes red, then purple, and finally blue. It is then mixed with some matter to give it a proper consistence, and subsequently introduced into moulds, where it dries in the form of rectangular cakes.

General Characters.—Litmus occurs in the form of rectangular cakes of from a quarter of an inch to an inch in length; they are light in weight, friable, finely granular, of an indigo-blue or deep violet colour, with a somewhat saline and pungent taste, and a violet-like odour. Litmus is reddened by acids, and restored to its original blue colour by alkalies; but it is not made green by alkalies, as is the case with most vegetable blue colours.

Official Preparations.

Blue Litmus Paper. | Red Litmus Paper.

Solution of Litmus.

These are official in the Appendix of the British Pharmacopœia for chemical testing.

---

2. CETRARIA ISLANDICA, *Ach.*

Iceland Moss.

(Bentley and Trimen's ' Medicinal Plants,' vol. iv. plate 302.)

Habitat.—This lichen is very abundant in high northern latitudes, where it may be found both on the mountains and in the plains. It is also commonly found in the mountainous parts of countries with a warmer climate, as in Britain and other parts of Europe, Asia, and North America. It is likewise met with in the Antarctic regions at Cape Horn.

Official Part and Name.—CETRARIA :—the dried lichen.

Cetraria.

Iceland Moss.

Synonym.—Iceland Lichen.

Name, Collection, and Commerce.—Its common name of Iceland Moss is altogether incorrect, for it is a lichen and not a moss, and, according to 'Pharmacographia,' none is exported from Iceland, our supplies appearing to be entirely derived from Sweden. A recent traveller in Iceland, Mr. Wright, states however, that it is exported to Europe, hence the produce of Iceland may reach us indirectly.

General Characters.—The thallus (*fig.* 57) is foliaceous, and much branched in an irregular dichotomous manner into fringed obtuse or truncate flattened lobes. It is crisp, somewhat cartilaginous, smooth, and usually brownish or greyish-white above, and whitish beneath, where it is marked at irregular intervals with small white depressed spots (*soredia*). The apothecia (which are but rarely present) are placed on the upper surface near the ends of the lobes, and appear as flat, more or less circular or shield-like bodies, of a dark rusty or chestnut-brown colour, and with raised borders (*fig.* 57, *a, a*). It is almost odourless when dry, but when moistened with water having a feeble sea-weed-like odour; its taste is mucilaginous and slightly bitter. A strong decoction gelatinises on cooling.

FIG. 57.—Thallus of *Cetraria islandica. a, a.* Apothecia.

Principal Constituents.—About 70 per cent. of *lichenin* or *lichen starch*, which is isomeric with starch and cellulose, and assumes a blue colour on the addition of iodine ; and from 2 to 3 per cent. of a bitter crystalline principle, termed *cetraric acid.*

Medicinal Properties.—Demulcent and slightly tonic; and when deprived of its bitter principle by maceration in a weak solution of an alkaline carbonate in cold water, it is nutritive.

Official Preparation.—Decoctum Cetrariæ.

ORDER 2.—FUNGI.

1. CLAVICEPS PURPUREA, *Tulasne.*

Ergot.

(Bentley and Trimen's ' Medicinal Plants,' vol. iv. plate 303.)

Habitat.—The sclerotioid condition of this fungus, known under the name of ergot, is not confined to rye, but is frequently found on wheat and barley, and on a large number of wild grasses; but as a general rule, the size and form of the ergot varies according to the species on which it occurs.

Official Part and Name.—ERGOTA :—the sclerotium of Claviceps purpurea, *Tulasne,* produced between the pales (*fig.* 59), and replacing the grain of Secale cereale, *Linn.*

Ergota.

Ergot.

Commerce and Names.—Our supplies are essentially imported from Vigo in Spain, and from Teneriffe; and in some degree also from France and Hamburg. It is commonly known as Ergot, Ergot of Rye, Spurred Rye, or Horned Rye.

General Characters.—Ergot (*figs.* 58 and 59) varies in length from one-third of an inch to an inch and a half, and from about one-twenty-fourth to one-third of an inch in diameter.

In form it is sub-cylindrical or obscurely triangular, tapering towards
the ends, generally arched or curved
somewhat like the spur of a cock,
hence its common name of spurred
rye ; it is longitudinally furrowed on
each side, but more especially on
that which is concave, and often
irregularly cracked. Its colour externally is violet-purple or purplish-
black, and internally whitish or
pinkish-white. It has a short frac-

FIG. 59.—A fully developed ergot, between the
pales of a flower of *Secale cereale.*

ture ; a peculiar and disagreeable
odour, more especially if the powder
be triturated with solution of potash ;
and a mawkish rancid taste.

Preservation. — Ergot is very
liable to become injured by keeping,
partly from the oxidation of its
contained oil, but more especially
from the ravages of a species of
mite, belonging to the genus *Trom-*

FIG. 58.—An ear of Rye (*Secale
cereale*) affected with ergot.

bidium ; it should therefore be thoroughly dried, kept in closed bottles in a dry place, and renewed once a year at least. Camphor has also been employed in preserving it ; and the addition of a few drops of chloroform has been found useful.

Principal Constituents.—Several peculiar principles have been found in ergot, but further experiments are still necessary before we can pronounce positively upon their therapeutical action. There is no *starch* in ergot. The principal constituents that have been indicated are about 30 per cent. of a *fixed oil*, about 4 per cent. of *sclerotic* or *sclerotinic acid*, *scleromucin*, *sclererythrin*, *scleroiodin*, *picrosclerotine*, *sclerocrystallin*, *scleroxanthin*, *ecboline* and *ergotine*, which are probably identical, and a crystalline alkaloid called *ergotinine*. So far as investigations enable us to judge, the active therapeutical constituents appear to be more especially sclerotic acid and scleromucin ; but Tanret believes that ergotinine is the essential active constituent. Kobert has recently stated that ergot contains three active principles, namely, *ergotinic acid*, *sphacelinic acid*, and an alkaloid *cornutine*.

Medicinal Properties.—In proper doses ergot acts principally on the muscular fibres of the uterus, causing their strong and continuous contraction, and hence it is a most valuable remedy in tedious parturition, and in the prevention of flooding after delivery. It may also be used generally as a hæmostatic, and it is also emmenagogue ; but in overdoses ergot is poisonous. *Dose.*—20 to 30 grains.

Official Preparation. —Extractum Ergotæ Liquidum. *Dose.*—10 to 30 minims.

This extract is used for the preparation of the official Ergotinum, which is described in the British Pharmacopœia as Purified Extract of Ergot, commonly called Ergotin, Ergotine, or Bonjean's Ergotine. *Dose.*—2 to 5 grains. Ergotinum is also employed for the preparation of Injectio Ergotini Hypodermica. *Dose, by subcutaneous in-*

iection.—3 to 10 minims.   The other preparations of Ergot are :—

> Infusum Ergotæ.   *Dose.*—1 to 2 fluid ounces.
> Tinctura Ergotæ.   *Dose.*—5 to 30 minims.

---

2. SACCHAROMYCES (TORULA, *Turpin*) CEREVISIÆ, *Meyen.*

Yeast Plant.

Official Product and Name.—CEREVISIÆ FERMENTUM : —the ferment obtained in brewing beer, and produced by Saccharomyces cerevisiæ, *Meyen.*

Cerevisiæ Fermentum.

Beer Yeast.

Nature.—The so-called Yeast Plant is a mycelial form of *Saccharomyces cerevisiæ* ; and the so-called Vinegar Plant is also a more developed form of the mycelium of the same fungus.

General Characters.—Beer Yeast is a viscid semi-fluid frothy mass, which, when examined by the microscope, is found to consist of an immense number of isolated roundish or oval cells (*bottom yeast*); or short branched filaments formed of united cells (*top yeast*).   It has a peculiar odour, and bitter taste.

Medicinal Properties.—When given internally it is tonic antiseptic, and laxative ; and when applied externally as a poultice, stimulant and antiseptic.   *Dose.*—½ to 1 ounce.

Official Preparation.—Cataplasma Fermenti.

PART II.

ANIMAL DRUGS.

UNDER the head of Animal Drugs we include all those which are derived from the Animal Kingdom, and which consist of animals with their products and educts, except those obtained as products of decomposition, which are official in the British Pharmacopœia, together with some others that are employed in the treatment of disease. These drugs, although few and unimportant in comparison with those derived from the Vegetable Kingdom, will be best described like them in the order of the natural historical relations of the animals from which they are obtained. We shall also take them in the same order as the vegetable drugs, by proceeding from those obtained from the highest animals down to those of the lowest development.

SUB-KINGDOM I.

VERTEBRATA.

CLASS I. MAMMALIA.

ORDER I.—RODENTIA.

CASTOR FIBER, *Linn.*

The Beaver.

Part Used and Name.—CASTOREUM:—the dried preputial follicles and their secretion, obtained from the Beaver, and separated from the somewhat shorter and smaller oil sacs which are frequently attached to them.

(*Not Official.*)

Castoreum.

Castor.

Commerce.—Castor is now exclusively obtained from the Hudson's Bay Territory, and is commonly known as North American, Canadian, or Hudson's Bay Castor.

General Characters.—The follicles are in pairs, usually about three inches long, and united by a ligamentous band; they are fig-shaped or somewhat pyriform, frequently wrinkled externally, brown, reddish-brown or greyish-black, firm, heavy, breaking with a resinous fracture; and containing a dry resinous reddish-brown or brown secretion, which has a very strong peculiar odour, and a bitter aromatic taste.

Principal Constituents.—*Volatile oil* containing traces of *carbolic acid* and *salicin, cholesterin, bitter resinous substance,* and a peculiar white crystalline fatty substance, termed *castorin.* The contents of the follicles should be, in great part, soluble in rectified spirit and in ether.

Medicinal Properties.—Stimulant and antispasmodic, in doses of from 5 to 10 grains; but rarely used at the present day.

Preparation.—The official preparation in the British Pharmacopœia of 1867 was Tinctura Castorei. *Dose.*—$\frac{1}{2}$ to 1 fluid drachm.

ORDER 2.—RUMINANTIA.

1. MOSCHUS MOSCHIFERUS, *Linn.*

The Musk Deer.

Official Product and Name.—MOSCHUS :—the dried secretion from the preputial follicles.

C C

Moschus.

Musk.

Varieties and Commerce.—The secretion of musk is peculiar to the male animal. Two varieties are more especially known in commerce, namely, China or Thibet Musk; and Russian, Siberian, or Kabarbine Musk. The former, which is imported from China and India, is the best variety, and is alone official.

General Characters.—Musk is in irregular somewhat unctuous grains of a dark reddish-brown or reddish-black colour, a very strong peculiar diffusible penetrating persistent odour, and a bitterish taste. It is contained in a roundish or oval sac, from about one and a half to two inches in diameter, which is nearly smooth on one side, and covered on the other or outer side by brownish-yellow or greyish adpressed bristle-like hairs, concentrically arranged around a nearly central orifice.

Test.—It should be free from earthy impurities. The ash should not exceed about 8 per cent.

Russian or *Siberian musk* is commonly in longer and flatter sacs or pods, and the contained secretion more compact, and of a less powerful and agreeable odour.

Substitutions and Adulterations.—Musk is very liable to adulteration—both the sacs and the contents being sometimes spurious; but these artificial musk pods, as also their contents, may be readily known by the absence of the characters given above of true musk.

Principal Constituents.—The most important constituent is an *odorous principle*, but this has not been isolated. A few drops of solution of potash added to musk increase its odour by setting free, as it is supposed, *ammonia*. Other constituents are *stearin, olein, cholesterin*, etc.

Medicinal Properties.—Stimulant and antispasmodic, like

castor; but it is little used at the present day.   It has also been regarded as aphrodisiac.

Dose.—5 to 10 grains.

---

2. OVIS ARIES, *Linn.*

The Sheep.

Official Product and Name.—SEVUM PRÆPARATUM :— the internal fat of the abdomen of the sheep; purified by melting and straining.   (*See also* Pepsin.)

Sevum Præparatum.

Prepared Suet.

Preparation.—Mutton suet is the fat from the neighbourhood of the kidneys of the sheep ; and the official suet is prepared by melting the fat over a slow fire in a water bath, and subsequent straining in order to separate the membranous portions.

General Characters. — Solid, smooth, white, almost odourless, and with a bland taste.   It is fusible at 103° F. (39°·4 C.).   It is entirely soluble in ether.

Principal Constituents.—It is principally composed of *stearin*; the remainder being *palmitin*, *olein*, and a trace of *hircin.*

Medicinal Properties.—It is used externally as an emollient.

Official Preparations.

Emplastrum Cantharidis. | Unguentum Hydrargyri.

Unguentum Hydrargyri is also used in the preparation of Linimentum Hydrargyri ; Suppositoria Hydrargyri; and Unguentum Hydrargyri Compositum.

3. BOS TAURUS, *Linn.*

The Ox and Cow.

Official Products and Names.—1. FEL BOVINUM PURI-
FICATUM :   the purified gall of the Ox.   2. LAC :—the fresh
milk of the Cow.   3. SACCHARUM LACTIS :—a crystallised
sugar, obtained from the whey of Milk by evaporation. (*See
also* Pepsin.)

1. Fel Bovinum Purificatum.

Purified Ox Bile.

Synonym.—Purified Ox Gall.

Preparation.—The gall bladder of the ox contains a
viscid greenish-brown alkaline liquid, with an unpleasant
odour, and a bitter but subsequent sweetish taste. When
purified, as directed in the pharmacopœia, by agitation with
rectified spirit, and inspissated to a pilular consistence, it
constitutes the official purified ox bile.

General Characters and Tests.—These are given in the
British Pharmacopœia as follows :—A yellowish-green substance, having a taste partly sweet and partly bitter, soluble
in water and in spirit.   A solution of one or two grains of it.
in about a fluid drachm of water, when treated, first with a
drop of freshly-made syrup consisting of one part of sugar
and four of water, and then with sulphuric acid cautiously
added until the precipitate at first formed is redissolved,
gradually acquires a cherry-red colour, which changes in
succession to carmine, purple, and violet.   Its watery solution gives no precipitate on the addition of rectified spirit.

Principal Constituents.—Bile, when separated from the
mucus of the gall bladder, as is the case with the official
prepared ox bile, contains *colouring, fatty,* and *biliary matters.*
The first gives to it its greenish colour ; the peculiar fatty
matter is *cholesterin,* which exists in healthy bile in but very

small proportion, but in certain morbid states of the liver it accumulates, and forms the chief constituent of gall stones. The biliary matter is composed of two acids in combination with soda, and termed *glycocholic* and *taurocholic acids.* The presence of these acids is demonstrated in ox bile by the change of colour, as given above in ' General Characters and Tests,' under the action of sulphuric acid and sucrose (sugar).

Medicinal Properties.—In small doses it is regarded as tonic, and in larger doses as a mild laxative. *Dose.*—5 to 10 grains.

2. Lac.

Milk.

General Characters.—Fresh milk, as obtained from the mammary glands of the cow, is an opaque, slightly alkaline, emulsive, white liquid, with a bland sweetish taste, and a faint peculiar odour.   Its specific gravity is usually given at about 1·030.

Principal Constituents.—It consists essentially of *butter* (*fat*), *casein* or *curd*, and *sugar of milk.*   When examined by the microscope, milk is seen to be composed of myriads of very minute globules, each of which is invested by a thin membrane, and the whole floating in a transparent liquid, which is thus rendered opaque.   The milk globules consist essentially of butter.   Sugar of milk being official, is described below.

Medicinal Properties.—Milk is eminently nutritious ; but medicinally it also acts as an emollient and demulcent, both when given internally, or locally applied.

Official Preparation.—Mistura Scammonii.

3. Saccharum Lactis.

Sugar of Milk.

$$C_{12}H_{24}O_{12}.$$

Preparation.—When the butter and casein have been removed from milk, the whey which remains is evaporated, and the sugar allowed to crystallise upon sticks or cords.

General Characters.—Usually in cylindrical masses, about two inches in diameter, with a cord or stick in the axis ; or in fragments of cakes.   It is greyish-white, crystalline on the surface and in its texture, translucent, hard, scentless, faintly sweet, and gritty when chewed.   It is soluble in about seven parts of water at common temperatures, and in about one part of boiling water.

Medicinal Properties.—It is demulcent and laxative ; but chiefly used as a vehicle for more active medicines.

Official Preparation. — Pulvis Elaterini Compositus. *Dose.*—$\frac{1}{2}$ grain to 5 grains.

Order 3.—PACHYDERMATA.

SUS SCROFA, *Linn.*

The Hog.

Official Products and Names.—1. Adeps Præparatus :— the purified fat of the hog. 2. Pepsin :—a preparation of the mucous lining of the fresh and healthy stomach of the pig, sheep, or calf.

1. Adeps Præparatus.

Prepared Lard.

Preparation.—Should be prepared as directed in the British Pharmacopœia.

General Characters and Tests.—A soft white fatty substance, melting at about 100° F. (37°·8 C.). It has no rancid odour; and dissolves entirely in ether. Distilled water in which it has been boiled, when cooled and filtered, gives no precipitate with nitrate of silver, and is not rendered blue by the addition of solution of iodine.

Principal Constituents.—Its principal constituent is *olein* (about 62 per cent.); the other constituents are *palmitin* and *stearin.*

Medicinal Properties.—Locally applied it is emollient; it is largely employed as a basis for ointments, etc.

Official Preparations.

Adeps Benzoatus.	Unguentum Hydrargyri Nitratis.
Emplastrum Cantharidis.	Unguentum Iodi.
Unguentum Hydrargyri.	Unguentum Terebinthinæ.

Adeps Benzoatus is also an ingredient in sixteen other ointments; and Unguentum Simplex, one of these preparations, is used in eight other ointments.

2. Pepsin.

Pepsin.

Preparation.—A form for its preparation is given in the British Pharmacopœia; and, as already noticed, it may be prepared from the mucous lining of the fresh and healthy stomach of the pig, sheep, or calf.

General Characters and Tests.—A light yellowish-brown powder, having a faint, but not disagreeable odour, and a slightly saline taste, without any indication of putrescence. Very little soluble in water or spirit. Two grains of it with an ounce of distilled water, to which five minims of hydrochloric acid have been added, form a mixture in which at least 100 grains of hard-boiled white of egg, passed through wire-gauze of 36 meshes per linear inch and made of No. 32 brass or copper wire, will dissolve on their being well mixed, digested, and well stirred together for thirty minutes at a temperature of 130° F. (54°·4 C.).

Medicinal Properties.—It is used largely as a substitute for the natural digestive fluid in dyspepsia, and more especially in atonic dyspepsia. *Dose.*—2 to 5 grains.

Order 4.—CETACEA.

PHYSETER MACROCEPHALUS, *Linn.*
The Sperm Whale.

Official Product and Name.—Cetaceum :—a concrete fatty substance, obtained, mixed with oil, from the head of the Sperm Whale. It is separated from the oil by filtration and pressure, and afterwards purified.

Cetaceum.
Spermaceti.

Purification and Commerce.—The Sperm Whale, from which spermaceti is obtained, inhabits the Pacific and Indian Oceans, and the *head matter*, as the above concrete fatty substance is commonly termed, is, as stated in the British Pharmacopœia, separated from the oil by filtration and pressure, and afterwards purified.

General Characters and Tests.—Spermaceti is crystalline, pearly-white, glistening, translucent, with little taste or odour, and reducible to powder by the addition of a little rectified spirit. It is insoluble in water, but soluble in ether, chloroform, or boiling rectified spirit. It is scarcely unctuous to the touch. Its melting point is from 111° to 122° F. (43°·9 to 50° C.) when tested by the method described in the 'General Characters and Tests' of Cera Flava (*see* Cera Flava).

Principal Constituents.—Spermaceti mainly consists of *Cetine* or *Palmitate of Cetyl.*

Medicinal Properties.—It was formerly given internally as a demulcent; but its use is now almost exclusively confined to its local application in the form of an emollient ointment.

Official Preparations.

Charta Epispastica. | Unguentum Cetacei.

CLASS II. AVES.

ORDER GALLINÆ.

GALLUS BANKIVA, var. DOMESTICUS, *Temminck.*

The Domestic Cock and Hen.

Official Products and Names.—1. OVI ALBUMEN :—the liquid white of the egg. 2. OVI VITELLUS :—the yolk of the egg.

Characters of the Egg.—The egg of the hen is too well known to need any particular notice; it consists of about 10 per cent. of shell with its lining membrane, 60 per cent. of albumen, and 30 per cent of yolk.

1. Ovi Albumen.

Egg Albumen.

General Characters.—The *albumen*, *glaire*, or *white of the egg*, is a transparent viscid glairy liquid, miscible with water, and is coagulated by heat at about 160° F., when it becomes white, opaque, and insoluble in water. It is coagulated by corrosive sublimate, and also by ether, in which latter character it differs from the albumen of blood serum. It is distinguished from casein by not being coagulated by acetic acid.

Medicinal Properties.—Used as an antidote in mineral poisoning, more especially in the cases of corrosive sublimate and sulphate of copper ; and also as a demulcent or sheathing agent in all cases of corrosive or acrid poisoning.

2. Ovi Vitellus.

Yolk of Egg.

General Characters and Composition.—Yolk or yelk of egg has a yellow colour and is coagulable by heat. It contains from about 48 to 55 per cent. of *water ;* about 14 per cent. of *vitellin* or *casein ;* about 30 per cent. of *fat ;* *various inorganic salts* to about 1·5 per cent. ; traces of *cholesterin, colouring matter,* etc.

Medicinal Properties.—Demulcent and emollient. It is chiefly used in the preparation of the official Mixture of French Brandy, which is used as a stimulant and nutritious mixture in exhausted states of the system.

Official Preparation.—Mistura Spiritus Vini Gallici.

CLASS III. PISCES.

ORDER I.—STURIONES.

ACIPENSER, *Linn.*

Sturgeon.

Official Product and Name.—ISINGLASS :—the swimming bladder or sound of various species of Acipenser, *Linn.,* prepared, and cut into fine shreds.

Ichthyocolla.

Isinglass.

Varieties and Sources.—There are several varieties of isinglass known in commerce. The Russian kinds are the most esteemed, but the Brazilian is also much used. Other varieties are New York, East Indian, Manila, and Hudson's Bay, but the sources of all except the Russian have not been accurately determined; the four latter varieties are, however, not obtained from species of *Acipenser*. A. Huso, *Linn.* (the Beluga); A. Güldenstadtii, *Brandt & Ratz.*, A. Ruthenus, *Linn.*, A. stellatus, *Pallas*, and probably others, are the sources of Russian isinglass.

General Characters.—Isinglass is whitish or yellowish in colour, light, semi-transparent, tasteless, and inodorous. It is insoluble in cold water, but readily soluble in boiling water, and forms a transparent jelly on cooling.

Isinglass is found in various forms, which are known as *purse, pipe, lump,* when formed of the dried unopened swimming bladder. Or when the bladder is opened, prepared, and dried, if unfolded, it constitutes *leaf* and *honeycomb isinglass*; or if folded, *staple* and *book isinglass*; or when rolled out, *ribbon isinglass*.

Principal Constituents.—Isinglass of fine quality consists of about 90 per cent. of *gelatine*; the remainder being about 8 per cent. of *albumen*, and some *earthy salts*.

Medicinal Properties.—Emollient and demulcent. But it is only placed in the Appendix of the British Pharmacopœia for the preparation of a test solution.

Official Preparation.—Solution of Isinglass. This preparation is placed in the Appendix of the British Pharmacopœia. It is especially used as a test to distinguish gallic from tannic acid; the latter giving with it a yellowish-white precipitate; the former has no effect upon it.

ORDER 2.—TELEOSTEÆ.

GADUS MORRHUA, *Linn.*

The Cod.

Official Product and Name.—OLEUM MORRHUÆ :—the oil extracted from the fresh liver of the cod, by the application of a heat not exceeding 180° F. (82°·2 C.).

Oleum Morrhuæ.

Cod-Liver Oil.

Synonym.—Oleum Jecoris Aselli.

Source.—Although *Gadus Morrhua* is the principal source of cod-liver oil, the commercial oil is also derived from other species of *Gadus.*

Preparation and Commerce.—The best oil is obtained by slowly heating the fresh livers to a temperature not exceeding 180° F., and then separating the oil which floats on the surface, and subsequent filtration. The average yield is about 40 per cent. It is prepared in enormous quantities in Newfoundland and Norway, and also in this country.

Varieties, General Characters, and Test.— Commercial oils vary much in colour according to their mode of preparation ; and are known from this circumstance as *pale-yellow*, *pale-brown*, and *dark-brown* ; they also vary in taste and odour. The *pale-yellow* is that which is prepared according to the directions of the British Pharmacopœia, and is alone, therefore, official. Its characters and test are given as follows :—Pale yellow, with a slight fishy odour, and bland fishy taste. A drop of sulphuric acid added to a few drops of the oil on a porcelain slab develops a violet colour, which soon passes to a yellowish or brownish-red.

Principal Constituents.—It contains about 80 per cent. of *olein*, 15 per cent. of *palmitin* and *stearin* ; small portions of *cholic* and *butyric acids*, and a substance called *gaduin* ; and minute traces of *iodine, bromine, phosphorus*, etc.

Medicinal Properties.—Cod-liver oil is a very digestible fat, and under favourable circumstances it fattens the patient and enriches the blood, and hence it is used to an enormous extent in phthisis, scrofulous affections, etc.

Dose.—1 to 8 fluid drachms.

SUB-KINGDOM II.

INVERTEBRATA.

CLASS I.   INSECTA.

ORDER I.—HYMENOPTERA.

APIS MELLIFICA, *Linn.*

The Hive Bee, or Honey Bee.

Official Products and Names.—1. MEL :—a saccharine secretion deposited in the honeycomb by Apis mellifica, *Linn.*   2. CERA FLAVA :—prepared from the honeycomb of the Hive Bee.   3. CERA ALBA :—yellow wax bleached by exposure to moisture, air, and light.

1. Mel.

Honey.

Production and Commerce.—Honey is secreted by flowers, from which it is sucked or lapped by the working or neuter bees by their tongue.   It is first deposited in a membranous bag below the middle of this organ, and then consigned to the *crop* or *honey-bag*, where it becomes slightly altered by the addition of an acid ; and ultimately it is disgorged and deposited in the honeycomb. Honey is obtained in this country and also imported. A hive which has never swarmed is considered to yield honey of the best quality ; this is called *virgin honey.*

General Characters.—When recently separated from the honeycomb, it is a viscid translucent liquid, of a light-yellowish or brownish-yellow colour, which gradually becomes partially crystalline and opaque. Its specific gravity, according to Stoddart, is 1·075. It has a peculiar somewhat aromatic odour, and a sweet characteristic taste. Honey, however, varies in its colour, odour, and taste, according to the age of the bees and the nature of the flowers on which they have fed.

Tests.—When boiled with water or five minutes and allowed to cool, it does not become blue with the solution of iodine. Incinerated it should not yield more than 0·2 per cent. ash, the solution of which in water acidulated with nitric acid should not afford more than a slight turbidity with solution of chloride of barium.

Principal Constituents.—Honey is essentially a *concentrated solution of sugar*, mixed with *waxy*, *odorous*, and *colouring matters*. According to Stoddart, the sugar of fresh honey is of three kinds : one crystallisable and analogous to *glucose* or *grape sugar* ; another is *uncrystallisable sucrose* (*lævulose*), or inverted sugar, and similar to the uncrystallisable brown syrup of the sugar cane ; and the third is *crystallised sucrose.* The proportions of these vary according to the age of the honey.

Medicinal Properties.—Honey is emollient, demulcent, and laxative ; it is also used as a vehicle for other medicines, both for internal and external use.

Official Preparations.—As honey is commonly more or less contaminated with organic impurities, which render it liable to ferment, it is directed in the British Pharmacopœia, before being used in medicine, to be clarified according to the process there given. Of this Clarified Honey, or Mel Depuratum, we have the following official preparations :—

Confectio Piperis.

Confectio Scammonii.

Confectio Terebinthinæ.

Mel Boracis.

Oxymel.

Oxymel Scillæ.

2. Cera Flava.

Yellow Wax.

Secretion and Preparation.—Yellow wax (*Bees' Wax*) is secreted by glands (*wax pockets*) placed on the ventral scales of the bee ; and is used by the insect to build the honey-comb, in which, as we have seen, the honey which they use as food is deposited. After the honey has been removed by draining and subsequent expression from the comb, the latter is melted in water, when the impurities subside, and the wax is decanted into moulds and left to cool.

General Characters and Tests.—Yellow wax is firm, breaks with a granular fracture, is not unctuous to the touch, has a yellowish colour, mild taste, and an agreeable honey-like odour. Should be readily and entirely soluble in hot oil of turpentine. Should not yield more than three per cent. to cold rectified spirit, and nothing to water or to a boiling solution of soda, the two latter fluids after filtration neither being turbid nor yielding a precipitate on the addition of hydrochloric acid. Specific gravity, 0·950 to 0·970. Melts at 146° F. (63°·3 C.), when tested in the following manner. Liquefy a few grains, and draw a little of the fluid up into a capillary tube ; fix a piece of the filled capillary tube to the bulb of a thermometer by thread ; immerse the bulb and tube in a beaker of water and heat the latter gently ; at the moment the opaque rod of wax becomes transparent, note the temperature. The solidifying point is two to three degrees lower than the melting point. Boiling water in which it has been agitated is not, when cooled, rendered blue by iodine.

3. Cera Alba.

White Wax.

Preparation.—White wax, as already stated, is yellow wax bleached by exposure to moisture, air, and light.

General Characters and Tests.—White wax is hard, nearly white, and translucent. It should respond to the tests for yellow wax. (*See* Cera Flava.)

Principal Constituents of Wax.—Yellow wax is essentially composed of *myricin* or *melissyl palmitate*, *cerotic acid* or *cerin*, and *cerolein*, in varying proportions in different specimens. White wax is essentially the same without the colouring matter of yellow wax.

Medicinal Properties of Wax.—Wax is demulcent when used internally ; but it is rarely used except externally as a mildly sheathing or protective agent, or as a basis for the application of other substances, as in many of the official ointments, plasters, etc.

Official Preparations.

1. Of CERA FLAVA :—

Cera alba.
Emplastrum Calefaciens.
Emplastrum Cantharidis.
Emplastrum Galbani.
Emplastrum Picis.
Emplastrum Saponis Fuscum.
Pilula Phosphori.

Unguentum Cantharidis.
Unguentum Hydrargyri Compositum.
Unguentum Picis Liquidæ.
Unguentum Resinæ.
Unguentum Sabinæ.
Unguentum Terebinthinæ.

2. Of CERA ALBA :—

Charta Epispastica.
Unguentum Cetacei.
Unguentum Simplex (which is also used in the preparation of eight other Unguenta).

ORDER 2.--HEMIPTERA.

COCCUS CACTI, *Linn.*

Cochineal.

Official Part and Name.—Coccus :—the dried female insect, reared on Opuntia cochinillifera, *Mills* ; and on other species of Opuntia.

Coccus.

Cochineal.

Production, Collection, and Commerce.—The cochineal insects (*fig.* 60, *a*, *b*, *c*, *d*) are especially reared for commercial purposes, in Mexico, Central America, and Teneriffe, on the Nopal plant (*Opuntia cochinillifera*), which is cultivated in plantations for their nourishment. The impregnated females (*fig.* 60, *d*) are placed on these plants to deposit their eggs ; young ones are soon developed, which are carefully reared, and when they have arrived at a proper degree of development, and the young wingless females have become fecundated and enlarged (*fig.* 60, *d*), the collection takes place, the insects being brushed off from the plants, and killed

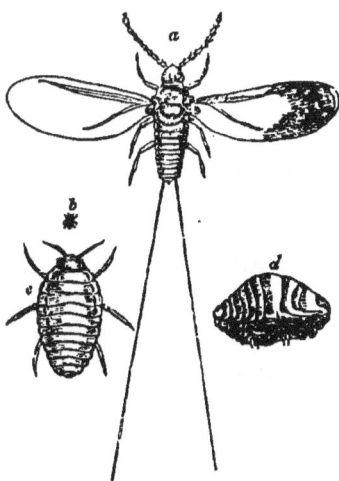

FIG. 60.—Cochineal insects (male and female). *a.* Male (magnified), with the wings expanded. *b.* Adult female (natural size). *c.* Adult female (magnified). *d.* Impregnated female (natural size).

by immersion in boiling water, or upon heated sheet-iron plates ; and then dried. In the first harvest, as it is called, in each year, the impregnated females are alone collected, and form the best cochineal ; but in the second harvest the young females are also collected. The female insect is alone of commercial value. Cochineal is exported chiefly from Mexico, Teneriffe, and Honduras. The male insect is much smaller than the female, and its body is terminated by two long diverging setæ (*fig.* 60, *a*).

General Characters.—Cochineal, as seen in commerce, is usually about one-fifth of an inch long, although some-times less ; somewhat oval in outline, flattish or concave beneath, convex above, transversely wrinkled, purplish-grey or purplish-black, and easily reduced to powder, which is dark red or puce-coloured. It has a very faint odour, and a slightly bitter taste.

Varieties.—Both the Honduras and Teneriffe commercial varieties of cochineal are distinguished into what are called the *silver* or *silvery grey* and *black* kinds. The silver grain, as it is also frequently termed, and the best when pure, has a purplish-grey colour owing to the presence of a whitish wool-like powder in its furrows and depressions ; and the latter, or *black grain*, is purplish-black. The difference between these two kinds appears to be due to the different ways in which they have been prepared. A third kind of cochineal is sometimes distinguished under the name of *Granilla* ; this is composed of small and of imperfectly developed insects, and is a very inferior variety.

Adulterations and Tests.—The *silver cochineal*, in order to increase its weight, has been first moistened with gum-water and then agitated with carbonate or sulphate of barium, or with carbonate of lead ; and the *black cochineal* has been similarly treated with ivory black or graphite. These adulterations may be readily detected by the pharmacopœia tests as follows :—When macerated in water no insoluble powder is separated. Ignited with free access of air, not much more than one per cent. of ash remains.

Principal Constituents.—The principal constituent is a colouring matter, which has been termed *cochinellin* or *car-*

mine. It is now generally called *carminic acid*, as it possesses acid properties.

Medicinal Properties.—Cochineal has been regarded as antispasmodic, anodyne, and diuretic ; but of this we have no reliable evidence. Its essential value is as a colouring agent.

Official Preparations.

Tinctura Cardamomi Composita.
Tinctura Cinchonæ Composita.
Tinctura Cocci.

ORDER 3.—COLEOPTERA.

CANTHARIS VESICATORIA, *De Geer.*

The Spanish Fly, or Blister Beetle.

Official Part and Name.—CANTHARIS :—the beetle dried.

Cantharis.

Cantharides.

Collection and Commerce.—These insects (*fig.* 61) are chiefly found upon species of *Oleaceæ*, as the ash, privet, lilac, and olive ; and of *Caprifoliaceæ*, as the elder and honeysuckle. They are collected by shaking the trees or shrubs, or beating the branches, when the insects fall upon cloths placed beneath the plants for that purpose, and are immediately killed with hot water, oil of turpentine, or vinegar, or by other means, and then dried.

Cantharides are principally imported from Hungary, Russia, and Sicily.

Preservation.—Cantharides are very liable to the attack of mites and other insects ; hence they should be preserved in well-stoppered bottles or other tightly closed vessels, in which a few drops of strong acetic acid, or a little camphor, or some other preservative substance, has also been placed.

General Characters.—From about three-quarters of an inch to an inch long, and a quarter of an inch broad (*fig.* 61, *a*, *b*, *c*, *d*), with two long elytra or wing-sheaths of a shining coppery-green colour, under which are two thin brownish transparent membranous wings (*fig.* 61, *b*). They have a strong and disagreeable odour ; and yield a greyish-brown powder, containing shining green particles.

Principal Constituents.—In addition to *odorous* and *fatty matters* these beetles contain a crystalline principle termed *cantharidin*, to which their active properties are due. It is

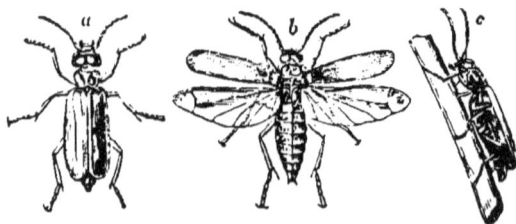

FIG. 61.--Cantharides.

contained in but small proportion, the amount has been variously estimated between 0·17 and 0·57, but some have obtained as much as 1 per cent., or even more.

Medicinal Properties.—A violent irritant, acting externally as a rubefacient and vesicant. When administered cantharides act especially upon the genito-urinary organs, and are diuretic and aphrodisiac, but their use internally requires great caution, as in overdoses, and in some cases even in moderate doses, they produce strangury. Cantharides in improper doses act as a powerful irritant poison. When given internally, the best form of administration is Tinctura Cantharidis ; the *dose* of which is from 5 to 20 minims.

Official Preparations.

Acetum Cantharidis.	is also used in the prepa-
Charta Epispastica.	ration of Collodium Vesi-
Emplastrum Calefaciens.	cans.
Emplastrum Cantharidis.	Tinctura Cantharidis.
Liquor Epispasticus, which	Unguentum Cantharidis.

Class II. ANNELIDA or ANNULOSA.

1. SANGUISUGA MEDICINALIS, *Savigny*.
The Speckled Leech.

2. SANGUISUGA OFFICINALIS, *Savigny*.
The Green Leech.   The Hungary Leech.

Official Part and Name.—Hirudo :—the Leech.

Hirudo.
The Leech.

Collection and Commerce.—Leeches are collected in various ways from the fresh-water ponds which they inhabit —that is, either by a net, or by the hand, or by baits, or by the gatherers walking into the ponds with naked feet, to which the leeches adhere. They are chiefly imported from Hamburg, but also from Bordeaux, Lisbon, and other parts of Europe.

Preservation.—Leeches are best preserved in an unglazed brown pan or glass vessel, filled for three or four inches from the bottom with a bed of pebbles, turf, moss, and some charcoal, and about two-thirds filled with clear river water or other soft water. They should be kept at a temperature from about 50° to 70° F., and not exposed to any sudden changes, and the water renewed at varying times, according to the temperature.

General Characters.—The body of the leech is soft, smooth, two or more inches long, and plano-convex, being rounded on the dorsal and flat on the ventral surface (*fig.* 62). It tapers towards each end, and is wrinkled in a transverse or annular manner by from about seventy to over a hundred rings of which it is composed. The back has an olive-green or somewhat blackish-green colour, with six rusty-red or somewhat yellowish-red longitudinal bands or stripes. At one end, the anterior (*fig.* 62, *a*), is the tri-radiate mouth, which is provided with three jaws, each of which is furnished with two rows of sharp teeth ; and the opposite extremity or posterior end is provided with a flattened muscular disc or sucker, by which the animal fixes itself (*fig.* 62, *b*).

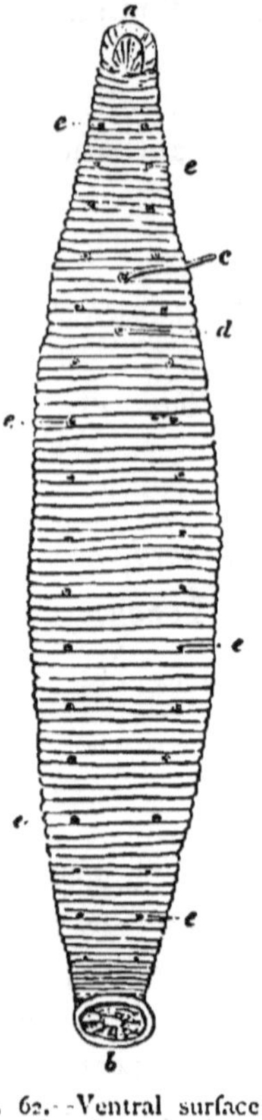

Sanguisuga medicinalis is distinguished by its greenish-yellow belly, spotted with black ; and the *Sanguisuga officinalis* by its olive-green not spotted belly.

FIG. 62.—Ventral surface of the Leech. *a.* Anterior disk or sucker *b.* Posterior disk. *c.* Penis. *d.* Vaginal orifice. *e, e, e, e, e, e.* Stigmata.

Uses.—Leeches are used for local abstraction of blood ; the ordinary quantity of blood drawn being from one to two fluid drachms, but by subsequent fomentation it may be increased to twice that amount. Before applying a leech the surface should be washed and thoroughly cleansed, and in some cases this is advantageously done with a little milk, or sugared water. The leech should be allowed to fill itself and drop away, and then further bleeding may be encouraged by hot fomentations.

Class III.   PORIFERA or SPONGIDA.

SPONGIA OFFICINALIS, *Linn.*
The Sponge.

Part Used and Name.—Spongia :—the dried skeleton of *Spongia officinalis*, and other species.

(*Not Official.*)

Spongia.
Sponge.

Collection and Preparation.—The best kinds are obtained by diving and cutting the sponges from the rocks below the sea to which they adhere ; the inferior sponges by tearing by a forked instrument or otherwise from the parts to which they cling.   The gelatinous matter, or flesh of the animal, is then removed by squeezing and washing, and the skeleton which remains dried.

General Characters and Tests.—Commercial sponge, as we have just seen, is the dried skeleton of the animal; it is formed of long branched anastomosing fibres, and is traversed by numerous canals, cavities, and pores.   When freed from stony concretions, sand, small shells, etc., it is soft, light, elastic, and in yellowish-brown or brown masses of various forms and sizes.   It evolves an animal odour when burned.   It absorbs water, and thereby swells up.   It is soluble in solution of potash.   Nitric acid colours it yellow.

Varieties and Commerce.—Two varieties are especially known in British commerce, Turkey and West Indian. *Turkey Sponge*, which is the best, is chiefly imported from Smyrna.   It occurs in cup-shaped masses, and has a finer texture than West Indian Sponge.   The common variety of this sponge is termed *honeycomb sponge*.   *West Indian*

Sponge is principally imported from the Bahama Islands, whence it is commonly known as *Bahama Sponge.* It has a more or less convex form with projecting lobes, and its component fibres are coarser, and have less cohesion than those of Turkey Sponge.

Principal Constituents. — The essential constituent of sponge is a proteid called *spongin* This is soluble in a hot solution of caustic potash.

Medicinal Properties.—Sponge when burned in a closed vessel, and powdered, was formerly official. Its efficacy is essentially due to iodine, which has almost or entirely superseded its use in medicine.

The uses of sponge economically, and for surgical purposes, are too familiar to need any description from us, and are, moreover, not within our province.

INDEX.

MEDICAL AND SCIENTIFIC WORKS

PUBLISHED BY MESSRS. LONGMANS & CO.

Ashby.—*NOTES ON PHYSIOLOGY FOR THE USE OF STUDENTS PREPARING FOR EXAMINATION.* With 120 Wood-cuts. By HENRY ASHBY, M.D. Lond. Fcp. 8vo. 5*s.*

Coats.—*A MANUAL OF PATHOLOGY.* By JOSEPH COATS, M.D. Pathologist to the Western Infirmary and the Sick Children's Hospital, Glasgow. With 339 Illustrations engraved on Wood. 8vo. 31*s.* 6*d.*

Cooke.—*TABLETS OF ANATOMY.* By THOMAS COOKE, F.R.C.S. Eng. B.A. B.Sc. M.D. Paris. Fourth Edition, being a selection of the Tablets believed to be most useful to Students generally. Post 4to. 7*s.* 6*d.*

Crookes.—*SELECT METHODS IN CHEMICAL ANALYSIS* (chiefly Inorganic). By WILLIAM CROOKES, F.R.S. V.P.C.S. With 37 Illustrations. 8vo. 24*s.*

Decaisne & Le Maout.—*A GENERAL SYSTEM OF BOTANY.* Translated from the French of E. LE MAOUT, M.D. and J. DECAISNE, by Mrs. HOOKER; with Additions by Sir J. D. HOOKER, C.B. F.R.S. Imp. 8vo. with 5,500 Woodcuts, 31*s.* 6*d.*

Dickinson.—*ON RENAL AND URINARY AFFECTIONS.* By W. HOWSHIP DICKINSON, M.D. Cantab. F.R.C.P. &c. With 12 Plates and 122 Woodcuts. 3 vols. 8vo. £3. 4*s.* 6*d.*

Erichsen.—*WORKS BY JOHN ERIC ERICHSEN, F.R.S.*

THE SCIENCE AND ART OF SURGERY: being a Treatise on Surgical Injuries, Diseases, and Operations. With 984 Illustrations. 2 vols. 8vo. 42*s.*

ON CONCUSSION OF THE SPINE, NERVOUS SHOCKS, and other Obscure Injuries of the Nervous System. Crown 8vo. 10*s.* 6*d.*

Ganot.—*WORKS BY PROFESSOR GANOT.* Translated by E. ATKINSON, Ph.D. F.C.S.

ELEMENTARY TREATISE ON PHYSICS, for the use of Colleges and Schools. With 5 Coloured Plates and 923 Wood-cuts. Large Crown 8vo. 15*s.*

NATURAL PHILOSOPHY FOR GENERAL READERS AND YOUNG PERSONS. With 2 Plates and 471 Woodcuts. Crown 8vo. 7*s.* 6*d.*

London: LONGMANS, GREEN, & CO.

Garrod.—*WORKS BY ALFRED BARING GARROD, M.D.*

A TREATISE ON GOUT AND RHEUMATIC GOUT (RHEU-MATOID ARTHRITIS). With 6 Plates, comprising 21 Figures (14 Coloured), and 27 Illustrations engraved on Wood. 8vo. 21s.

THE ESSENTIALS OF MATERIA MEDICA AND THERA-PEUTICS. New Edition, revised and adapted to the New Edition of the British Pharmacopœia, by NESTOR TIRARD, M.D. Crown 8vo. 12s. 6d.

Gray.—*ANATOMY, DESCRIPTIVE AND SURGICAL.* By HENRY GRAY, F.R.S. late Lecturer on Anatomy at St. George's Hospital. With 569 Woodcut Illustrations, a large number of which are coloured. Re-edited by T. PICKERING PICK, Surgeon to St. George's Hospital. Royal 8vo. 36s.

Hassall.—*THE INHALATION TREATMENT OF DISEASES OF THE ORGANS OF RESPIRATION,* including Consumption. By ARTHUR HILL HASSALL, M.D. With 19 Illustrations of Apparatus. Crown 8vo. 12s. 6d.

Helmholtz.—*WORKS BY PROFESSOR HELMHOLTZ.*

ON THE SENSATIONS OF TONE AS A PHYSIOLOGICAL BASIS FOR THE THEORY OF MUSIC. Translated by A. J. ELLIS, F.R.S. Royal 8vo. 28s.

POPULAR LECTURES ON SCIENTIFIC SUBJECTS. Translated and Edited by EDMUND ATKINSON, Ph.D. F.C.S. With a Preface by Professor Tyndall, F.R.S. and 68 Woodcuts. 2 vols. crown 8vo. 15s.; or separately, 7s. 6d.

Herschel.—*OUTLINES OF ASTRONOMY.* By Sir J. F. W. HERSCHEL, Bart. M.A. With Plates and Diagrams. Square crown 8vo. 12s.

Hewitt.—*THE DIAGNOSIS AND TREATMENT OF DIS-EASES OF WOMEN, INCLUDING THE DIAGNOSIS OF PREG-NANCY.* By GRAILY HEWITT, M.D. New Edition, in great part re-written and much enlarged, with 211 Engravings on Wood, of which 79 are new in this Edition. 8vo. 24s.

Holmes.—*A SYSTEM OF SURGERY,* Theoretical and Practical, in Treatises by Various Authors. Edited by TIMOTHY HOLMES, M.A. and J. W. HULKE, F.R.S. 3 vols. royal 8vo. £4. 4s.

Jones.—*THE HEALTH OF THE SENSES: SIGHT, HEARING, VOICE, SMELL AND TASTE, SKIN;* with Hints on Health, Diet, Education, Health Resorts of Europe, &c. By H. MACNAUGHTON JONES, M D. Crown 8vo. 3s. 6d.

London : LONGMANS, GREEN, & CO

Kolbe.—*A SHORT TEXT-BOOK OF INORGANIC CHEMISTRY.*
By Dr. HERMANN KOLBE. Translated from the German by
T. S. HUMPIDGE, Ph.D. With a Coloured Table of Spectra
and 66 Illustrations. Crown 8vo. 7s. 6d.

Liveing.—*WORKS BY ROBERT LIVEING, M.A. M.D.*

HANDBOOK ON DISEASES OF THE SKIN. With especial
reference to Diagnosis and Treatment. Fcp. 8vo. 5s.

NOTES ON THE TREATMENT OF SKIN DISEASES.
18mo. 3s.

Longmore. — *GUNSHOT INJURIES;* their History,
Characteristic Features, Complications, and General Treatment.
By Surgeon-General Sir T. LONGMORE, C.B., F.R.C.S. With
58 Illustrations. 8vo. 31s. 6d.

Miller.—*WORKS BY W. ALLEN MILLER, M.D. LL.D.*

THE ELEMENTS OF CHEMISTRY, Theoretical and Practical. Re-edited, with Additions, by H. MACLEOD, F.C.S.
3 vols. 8vo.
>> PART I.—CHEMICAL PHYSICS, 16s.
>> PART II.—INORGANIC CHEMISTRY, 24s.
>> PART III.—ORGANIC CHEMISTRY, 31s. 6d.

*AN INTRODUCTION TO THE STUDY OF INORGANIC
CHEMISTRY.* With 71 Woodcuts. Fcp. 8vo. 3s. 6d.

Murchison.—*WORKS BY CHARLES MURCHISON, M.D.*

*A TREATISE ON THE CONTINUED FEVERS OF GREAT
BRITAIN.* Revised by W. CAYLEY, M.D. Physician to the
Middlesex Hospital. 8vo. with numerous Illustrations, 25s.

*CLINICAL LECTURES ON DISEASES OF THE LIVER,
JAUNDICE, AND ABDOMINAL DROPSY.* Revised by T. LAUDER
BRUNTON, M.D. and Sir JOSEPH FAYRER, M.D. 8vo. with
43 Illustrations, 24s.

Owen.—*THE COMPARATIVE ANATOMY AND PHYSIOLOGY
OF THE VERTEBRATE ANIMALS.* By Sir RICHARD OWEN,
K.C.B. &c. With 1,472 Woodcuts. 3 vols. 8vo. £3. 13s. 6d.

Paget.—*WORKS BY SIR JAMES PAGET, BART. F.R.S.*

CLINICAL LECTURES AND ESSAYS. Edited by F.
HOWARD MARSH, Assistant-Surgeon to St. Bartholomew's
Hospital. 8vo. 15s.

LECTURES ON SURGICAL PATHOLOGY. Re-edited by the
AUTHOR and W. TURNER, M.B. 8vo. with 131 Woodcuts, 21s.

London : LONGMANS, GREEN, & CO.

Payen.—*INDUSTRIAL CHEMISTRY;* a Manual for Manufacturers, and for Colleges or Technical Schools; a Translation of PAYEN's 'Précis de Chimie Industrielle.' Edited by B. H. PAUL. With 698 Woodcuts. Medium 8vo. 42*s*.

Quain.—*A DICTIONARY OF MEDICINE.* By Various Writers. Edited by R. QUAIN, M.D. F.R.S. &c. With 138 Woodcuts. Medium 8vo. 31*s*. 6*d*. cloth, or 40*s*. half-russia; to be had also in 2 vols. 34*s*. cloth.

Quain's Elements of Anatomy.—The Ninth Edition. Re-edited by ALLEN THOMSON, M.D. LL.D. F.R.S. S.L. & E. EDWARD ALBERT SCHÄFER, F.R.S. and GEORGE DANCER THANE. With upwards of 1,000 Illustrations engraved on Wood, of which many are Coloured. 2 vols. 8vo. 18*s*. each.

Schäfer.—*THE ESSENTIALS OF HISTOLOGY, DESCRIPTIVE AND PRACTICAL.* For the use of Students. By E. A. Schäfer, F.R.S. With 281 Illustrations. 8vo. 6*s*. or Interleaved with Drawing Paper, 8*s*. 6*d*.

Schellen.—*SPECTRUM ANALYSIS IN ITS APPLICATION TO TERRESTRIAL SUBSTANCES,* and the Physical Constitution of the Heavenly Bodies. By Dr. H. SCHELLEN. Translated by JANE and CAROLINE LASSELL. Edited by Capt. W. DE W. ABNEY. With 14 Plates (including Angström's and Cornu's Maps) and 291 Woodcuts. 8vo. 31*s*. 6*d*.

Smith, H. F.—*THE HANDBOOK FOR MIDWIVES.* By HENRY FLY SMITH, M.B. Oxon. M.R.C.S. late Assistant-Surgeon at the Hospital for Sick Women, Soho Square. With 41 Woodcuts. Crown 8vo. 5*s*.

Smith, T.—*A MANUAL OF OPERATIVE SURGERY ON THE DEAD BODY.* By THOMAS SMITH, Surgeon to St. Bartholomew's Hospital. A New Edition, re-edited by W. J. WALSHAM. With 46 Illustrations. 8vo. 12*s*.

Thomson's Conspectus. — Adapted to the British Pharmacopœia of 1885. Edited by NESTOR TIRARD, M.D. Lond. F.R.C.P. New Edition, with an Appendix containing notices of some of the more important non-official medicines and preparations. 18mo. 6*s*.

West.—*WORKS BY CHARLES WEST, M.D. &c.*

LECTURES ON THE DISEASES OF INFANCY AND CHILD-HOOD. 8vo. 18*s*.

THE MOTHER'S MANUAL OF CHILDREN'S DISEASES. Crown 8vo. 2*s*. 6*d*.

London : LONGMANS, GREEN, & CO.

www.ingramcontent.com/pod-product-compliance
Lightning Source LLC
Chambersburg PA
CBHW021343210326
41599CB00011B/732